Climate Justice

Climate Justice

Integrating Economics and Philosophy

Edited by
Ravi Kanbur and Henry Shue

UNIVERSITY PRESS

OXFORD
UNIVERSITY PRESS

Great Clarendon Street, Oxford, OX2 6DP,
United Kingdom

Oxford University Press is a department of the University of Oxford.
It furthers the University's objective of excellence in research, scholarship,
and education by publishing worldwide. Oxford is a registered trade mark of
Oxford University Press in the UK and in certain other countries

First Edition published in 2019

Impression: 1

Published in the United States of America by Oxford University Press
198 Madison Avenue, New York, NY 10016, United States of America

British Library Cataloguing in Publication Data

Data available

Library of Congress Control Number: 2018939909

ISBN 978–0–19–881324–8

Printed and bound in Great Britain by
Clays Ltd, Elcograf S.p.A.

Contents

List of Figures

List of Tables

List of Contributors

Antonio Bento is a Professor at the Sol Price School of Public Policy and the Department of Economics of the University of Southern California. He is also a research associate of the National Bureau of Economic Research, and a research fellow of the Schwarzenegger Institute for State and Global Policy. Professor Bento is an applied microeconomist with a research program in the areas of environmental, energy, urban, and public economics. His work has been published in leading journals including the *American Economic Review*, the *Journal of Environmental Economics and Management*, the *Journal of Urban Economics*, and the *Energy Journal*. Professor Bento contributed to the New York State Climate Change Action Plan, the New York State Biofuels Roadmap, the UN Scientific Committee on Problems of the Environment, served as a contributing author to the Intergovernmental Panel on Climate Change Fifth Assessment Report, and was recently appointed as a lead author to the International Panel on Social Progress.

Simon Caney is Professor in Political Theory at the University of Warwick. He works on issues in contemporary political philosophy, and focuses in particular on issues of environmental, global, and intergenerational justice. He is completing two books—*Global Justice and Climate Change* (with Derek Bell) and *On Cosmopolitanism*—both of which are under contract with Oxford University Press. He is the author of *Justice Beyond Borders* (Oxford University Press, 2005). He co-edited *Climate Ethics: Essential Readings* (Oxford University Press, 2010) with Stephen Gardiner, Dale Jamieson, and Henry Shue. His publications have appeared in journals such as *Philosophy & Public Affairs*, *Journal of Political Philosophy*, and *Politics, Economics and Philosophy*. He has also written reports for public bodies, including a background paper on 'Ethics and Climate Change' for the World Bank, and a report on 'Climate Change, Equity and Stranded Fossil Fuel Assets' for Oxfam USA.

Ottmar Edenhofer is Designated Director of the Potsdam Institute for Climate Impact Research and Professor for the Economics of Climate Change at the Technical University Berlin. Moreover, he is founding director of the Mercator Research Institute on Global Commons and Climate Change as well as adviser to the World Bank regarding issues of economic growth and climate protection. He is a member of the German National Academy of Sciences (Leopoldina) and of the National Academy of Science and Engineering (acatech). He has published numerous articles in leading peer-reviewed journals and high-ranking disciplinary journals and authored various books. From 2008 to 2015 he served as Co-Chair of Working Group III of the Nobel Laureate Intergovernmental Panel on Climate Change, shaping the Fifth Assessment Report on

Climate Change Mitigation substantially, which has been considered a landmark report and provided the scientific basis for the Paris Agreement.

Stéphane Hallegatte is the Lead Economist with the Global Facility for Disaster Reduction and Recovery (GFDRR) at the World Bank. He joined the World Bank in 2012 after ten years of academic research. He was a lead author of the Fifth Assessment Report of the Intergovernmental Panel on Climate Change. He is the author of dozens of articles published in international journals in multiple disciplines and of several books. He also led several World Bank reports including *Shock Waves* in 2015, and *Unbreakable* in 2016. He was the team leader for the *World Bank Group Climate Change Action Plan*, a large internal coordination exercise to determine and explain how the Group will support countries in their implementation of the Paris Agreement. He holds an engineering degree from the École Polytechnique (Paris) and a PhD in economics from the École des Hautes Études en Sciences Sociales (Paris).

Nicole Hassoun is Associate Professor in Philosophy at Binghamton University and a visiting scholar at Cornell University. She has published more than seventy papers in journals like the *American Philosophical Quarterly, Journal of Development Economics, Journal of Applied Ethics, American Journal of Bioethics*, and *Philosophy and Economics*. Her first book, *Globalization and Global Justice: Shrinking Distance, Expanding Obligations*, was published by Cambridge University Press in 2012 and her manuscript *Global Health Impact: Extending Access on Essential Medicines for the Poor* is under contract with Oxford University Press. Professor Hassoun also heads the Global Health Impact project intended to extend access to medicines to the global poor. Global Health Impact launched at the World Health Organization in January 2015. The project is intended to assist policymakers in setting targets for and evaluating efforts to increase access to essential medicines.

Anders Herlitz, PhD, is a researcher at the Institute for Futures Studies in Stockholm where he works within the research program Climate Ethics and Future Generations and leads the research project Good and Just Allocation of Health-Related Resources. Dr. Herlitz's work primarily focuses on value theory, distributive ethics, and population-level bioethics.

Michael Jakob is a researcher at the Mercator Research Institute on Global Commons and Climate Change (MCC) in Berlin. Before joining the MCC, he spent more than five years as a PhD student and postdoc researcher at the Potsdam Institute for Climate Impact Research. He holds a PhD in economics from the Technical University of Berlin and has obtained degrees in physics, economics, and international relations from universities in Munich, St. Gallen, and Geneva. His research interests include climate change mitigation in developing countries, the political economy of climate policy, as well as the interlinkages between environmental policy and human well-being.

Ravi Kanbur is T. H. Lee Professor of World Affairs, International Professor of Applied Economics and Management, and Professor of Economics at Cornell University. He has served on the senior staff of the World Bank including as Chief Economist for Africa. He has also published in the leading economics journals, including the *Journal of Political Economy, American Economic Review, Review of Economic Studies, Journal of*

Economic Theory, and *Economic Journal*. He is President of the Human Development and Capabilities Association, Chair of the Board of United Nations University-World Institute for Development Economics Research, Co-Chair of the Scientific Council of the International Panel on Social Progress, member of the OECD High Level Expert Group on the Measurement of Economic Performance, past president of the Society for the Study of Economic Inequality, past member of the High Level Advisory Council of the Climate Justice Dialogue, and past member of the Core Group of the Commission on Global Poverty.

Anja Karnein is Associate Professor of Philosophy at Binghamton University (SUNY) where she teaches political philosophy and ethics. Her primary research interests concern questions of climate justice and our duties to future generations. Since 2013 she has been editor of the domain "Climate, Nature, Ethics" of the interdisciplinary journal *WIREs Climate Change*.

J. Paul Kelleher is Associate Professor of Bioethics and Philosophy at the University of Wisconsin-Madison. He studies ethical and philosophical aspects of public policy, especially health policy and climate change policy.

Ulrike Kornek received her PhD in economics from the Technische Universität Berlin in 2015. After working at the Potsdam Institute for Climate Impact Research, she currently holds a position as a post-doctoral researcher at the Mercator Research Institute on Global Commons and Climate Change. Her research interests are in environmental economics with a focus on international climate policy using numerical and analytical modeling. She also holds a diploma in physics from the Otto-von-Guericke Universität in Magdeburg, Germany.

Alex Lenferna is a Fulbright Scholar and PhC in the University of Washington Philosophy Department, and an Endeavour Research Fellow at the University of New South Wales' Practical Justice Initiative Climate Justice Research Stream. His research focuses on climate ethics and his PhD dissertation makes a moral case for a rapid and just transition away from fossil fuels. Alex holds master's degrees in philosophy from Rhodes University and the University of Washington, completing the former as a Mandela Rhodes Scholar. He received graduate certificates in both environmental studies and climate science from the Universities of Kansas and Washington respectively. Alex has recently served as a research consultant with 350.org, a fellow with Carbon Washington, a leader on a number of divestment campaigns, and a climate justice steward with the Alliance for Jobs and Clean Energy. He is a first-generation South African whose family hails from Mauritius.

Dominic Lenzi is a post-doctoral researcher at the Mercator Research Institute on Global Commons and Climate Change, in the working group Scientific Assessments, Ethics, and Public Policy. He is a philosopher working on ethical and political issues raised by climate change and the sustainability of the global commons. Dominic's current research focuses upon just entitlements to natural resources, and the ethics of negative emissions.

Jan Minx, PhD, is head of the MCC Working Group Applied Sustainability Science and Professor for Climate Change and Public Policy at the Priestley International Centre for

Climate at the University of Leeds. He takes an interdisciplinary research approach to issues of energy, climate change, and sustainable development with applications to cities and infrastructure, global supply chain networks, as well as historic and future transformation processes of societies and their governance. Before joining the MCC, he was Head of the Technical Support Unit of the Working Group on 'Mitigation of Climate Change' of the Intergovernmental Panel on Climate Change (IPCC) coordinating the Working Group's contribution to the IPCC's Fifth Assessment Report. He has co-authored articles in scientific journals such as *Science, Nature, Climate Change,* and the *Proceedings of the National Academy of Sciences* among others and is a member of the editorial board of the journals *Economic Systems Research* and *Carbon Balance and Management.*

Julie A. Nelson is Professor of Economics at the University of Massachusetts Boston and Senior Research Fellow at the Global Development and Environment Institute at Tufts University. Her research interests include ecological economics, ethics and economics, and feminist economics. She is the author of many publications including *Economics for Humans* (University of Chicago Press, 2nd edition, 2018) and articles in journals ranging from *Econometrica* and *Ecological Economics* to *Economics and Philosophy* and *Hypatia: Journal of Feminist Philosophy.* She is editor of the Economics and Business Ethics section of the *Journal of Business Ethics,* and is the 2018 President-Elect of the Association for Social Economics.

John Nolt is Distinguished Service Professor in the Department of Philosophy at the University of Tennessee and a Research Fellow in the Energy and Environmental Policy Program at the Howard H. Baker Jr. Center for Public Policy. His main areas of research are philosophical logic, formal value theory, environmental ethics, intergenerational ethics, and climate ethics. He is the author or editor of seven books, three on logic and four on environmental ethics—most recently *Environmental Ethics for the Long Term* (Routledge, 2015). Nolt is a long-time environmental activist and proponent of sustainable living, and for many years he chaired the Committee on the Campus Environment, which advises the Chancellor at the University of Tennessee on campus environmental policy.

Eugen Pissarskoi is a member of the departments of Ethics and Education and Nature and Sustainable Development at the International Centre for Ethics in the Sciences and Humanities of the Eberhard Karls University of Tübingen. Having studied philosophy and economics at the University of Mannheim and the Free University of Berlin, he received his PhD from the Institute for Philosophy, Free University of Berlin, with the thesis "Climate Change and Social Welfare." This has been published as *Gesellschaftliche Wohlfahrt und Klimawandel* (2014) with oekom verlag. From 2010 to 2017, he was researcher at the Institute for Ecological Economy Research, Berlin, and he worked with the research group "Limits and Objectivity of Scientific Foreknowledge: The Case of Energy Outlooks" at Karlsruhe Institute of Technology in 2011. In his research, he mainly focuses on decision-making under uncertainty, methodology of applied ethics, and environmental ethics.

Matthew Rendall is Lecturer in Politics and International Relations at the University of Nottingham. Much of his work has focused on large-scale war and peace, including such topics as the ethics of nuclear deterrence, whether there is a "separate peace" among democracies, and collective security. He has a long-standing interest in Russian foreign policy, particularly during the period of the Concert of Europe (1815–54). More recently he has also been writing about intergenerational justice, climate change, and other topics in moral philosophy and political theory.

Adam Rose is a Research Professor in the University of Southern California (USC) Sol Price School of Public Policy, and a Fellow of the Schwarzenegger Institute for State and Global Policy. Professor Rose's primary research interest is the economics of disasters. He has spearheaded the development of a comprehensive economic consequence analysis framework and has done pioneering research on resilience. His other major research area is the economics of energy and climate change policy, where he has focused on the aggregate and distributional impacts of climate mitigation policy, including advancing methodologies in both computable general equilibrium, and macroeconometric modeling. Professor Rose is the author of several books and 250 professional papers, including most recently *Defining and Measuring Economic Resilience from a Societal, Environmental and Security Perspective* (Springer), *Economic Consequence Analysis Tool* (Springer), and *The Economics of Climate Change Policy* (Edward Elgar). He was recently elected President of the International Society for Integrated Risk Management and a Fellow of the Regional Science Association International.

Julie Rozenberg is an Economist with the World Bank Sustainable Development Group. Her work includes green growth, climate change mitigation strategies, climate change adaptation, and disaster risk management. She is an author of "Decarbonizing Development: Three Steps to a Zero Carbon Future" and "Shockwaves: Managing the Impacts of Climate Change on Poverty." She focuses on helping World Bank teams and clients take climate change constraints and other long-term uncertainties into account in the preparation of projects and strategies, in order to build resilience in World Bank client countries. She holds an engineering degree from the École Nationale Supérieure de Techniques Avancées and a PhD in economics from the École des Hautes Études en Sciences Sociales in Paris.

Henry Shue was the co-founder of the Institute for Philosophy and Public Policy at the University of Maryland (1976–87), the inaugural Hutchinson Professor of Ethics and Public Life at Cornell University (1987–2002), and Professor of International Relations, University of Oxford (2002–7). Best known for *Basic Rights: Subsistence, Affluence, and U.S. Foreign Policy* (Princeton, 1980; 2nd edition, 1996), he published his first two decades of writings on climate change as *Climate Justice: Vulnerability and Protection* (Oxford University Press, 2014). His articles on the morality of violence appeared as *Fighting Hurt: Rule and Exception in Torture and War* (Oxford University Press, 2016). He is currently writing a series of articles on the urgency of action on climate change in light of duties of justice to future generations.

Dan Wei is a Research Associate Professor in the Sol Price School of Public Policy of USC. Her research focuses on economic modeling of energy and climate change policies and economic consequence of natural or man-made disasters. She performed macroeconomic impact analyses of state climate action plans for several major states in the US. She contributed to the capacity building on macroeconomic impact modeling of climate action plans in Guangdong, China and Baja California, Mexico. In the area of economics of disasters, she made significant contributions in economic impact studies of major disaster scenarios for the United States Geological Survey. She was a key investigator in an National Science Foundation study to analyze dynamic economic resilience of businesses to disasters and a Federal Emergency Management Agency study to develop a disaster deductible/credit formula for post-disaster public assistance. She recently led a study to develop and apply an economic framework to evaluate resilience in recovering from major port disruptions.

1

Climate Justice: Integrating Economics and Philosophy

Ravi Kanbur and Henry Shue

1.1. Introduction

Climate justice requires sharing the burdens and benefits of climate change and its resolution equitably and fairly. It brings together justice between generations and justice within generations. In particular it requires that attempts to address justice between generations through various interventions designed to curb greenhouse emissions today do not end up creating injustice in our time by hurting the currently poor and vulnerable. More generally, issues of distribution and justice are of paramount importance in any discourse on climate change. The United Nations Sustainable Development Goals (SDGs) summit in September 2015, and the Conference of Parties (COP) to the Framework Convention on Climate Change in Paris in December 2015, brought climate change and its development impact, including climate justice, center stage in global discussions.

In the run-up to Paris, Mary Robinson, former president of Ireland and the UN Secretary General's Special Envoy for Climate Change, instituted the Climate Justice Dialogue.[1] The objectives of the Climate Justice Dialogue were "to mobilise political will and creative thinking to shape an ambitious and just international climate agreement in 2015." As part of this project, the global High Level Advisory Committee to the Climate Justice Dialogue was formed in 2013. The Committee included former presidents and other leaders from the fields of politics, science, business, civil society, and academia, and publicly advocated for climate justice through their own work and via platforms

[1] <http://www.mrfcj.org/our-work/climate-justice-dialogue/>.

facilitated by the Mary Robinson Foundation. Together, they signed the Declaration on Climate Justice, which is reproduced as an Appendix to this volume.

Two of the academic disciplines which have engaged climate change and its justice dimensions are economics and philosophy. We, an economist and a philosopher, served on the High Level Advisory Committee of the Climate Justice Dialogue. Over the three years of our service on the committee, we noted the overlap and mutual enforcement between the economic and philosophical discourses on climate justice. But we also noted the great need for these strands to come together to support the public and policy discourse, which does not—and cannot afford to—see things in separate silos. This volume is the result.

In the broad policy and public discourse, "climate justice" has now become a catch-all term for the intersection between patterns and paths of climate change on the one hand, and their consequences for different conceptions of justice on the other. There are many different formulations of climate change, including the claim that it does not exist at all. And there are even more perspectives on justice, with each conceptualization and specification being deeply contested. Since climate justice is at the intersection of these terrains, it is not surprising that it is also a topic where many academic disciplines seek to make their contributions to a dominant policy issue of our time. Outside of the natural sciences, philosophy and economics are two disciplines central to the climate change discourse.

This volume has contributions from economists and philosophers who are clearly aware of, and influenced by, the other discipline's contributions. In this overview we attempt to characterize the various approaches, how they overlap and interact, and what they have already learned from each other and might still have to learn. Section 1.2 provides a brief narrative of the overlapping perspectives of economics and philosophy as seen through the chapters in this volume. Section 1.3 presents a chapter by chapter account of the volume. Section 1.4 concludes with a tentative exploration of some paths to better integration of the two disciplinary approaches to climate justice.

1.2. Overlapping Perspectives

One way to frame how different disciplines can engage with climate justice and the impact of alternative policies on climate justice, is to start with a distinction between the positive and the normative. A first step in evaluating any policy, which includes the policy of doing nothing, or continuing with the "business-as-usual" scenario of the current policy, is to lay out all the implications of that policy from now into the future. A second step is to evaluate the normative implications of alternative policy paths. Of course the two are not unrelated, because the effort we put into elucidating and elaborating this or that implication of a policy will depend upon how important we think that implication is for the normative evaluation.

The natural sciences are paramount in developing the implications of anthropogenic activities like greenhouse gas emissions on the climate. But the implications of climate change for human and non-human well-being bring us to the social sciences since it involves the organization and response of economic and social relations to climate change. Here economic modeling is central in predicting time paths of income and other economic variables through a range of Integrated Assessment Models (IAMs) as seen in the famous *Stern Review* (Stern, 2007) and many other exercises. These exercises are contested on the modeling, but they are the key input into the next step, of evaluating the time path generated by different policies.

The specifics and details of IAMs appear to be the preserve of economists, but it is in the evaluation of the alternative time paths generated by these models that economists and philosophers overlap in their perspectives and have much to learn from each other, and further integration between the two has a lot to contribute to the policy discourse. The potential overlaps, and debates, can in turn be put into two broad categories: (i) evaluation within a time period or generation; (ii) evaluation across time periods or generations. Cutting across these are issues related to focusing on economic outcomes versus a broader range of considerations going beyond the narrowly economic.

Coming then to normative evaluation within a time period or generation, one feature of note is that the Pareto criterion of economic textbooks, that in comparing two scenarios one is pronounced to be superior if in it no one is worse off and at least one person is better off compared to the other, is difficult to find in economic analyses of climate change. Thus it is recognized that there are gainers and losers from climate policy, within generations (and across generations). The usual trope that economics ignores distribution is not found in climate change analysis, as argued in Kanbur (2015). And it is not found in the economics or philosophy chapters in this volume. Rather, Rozenberg and Hallegatte (Chapter 2) project the consequences of different scenarios for poverty in the year 2030. Rose, Wei, and Bento (Chapter 4) assess the income distribution implications of emission reductions proposals in the Paris Agreement according to different equity criteria. Jakob et al. (Chapter 3) assess the equity implications of alternative uses of the revenues from a carbon tax or other forms of taxation of natural resource rents. Philosophers Hassoun and Herlitz (Chapter 5) focus on the issue of within-country and between-country inequality.

In all of these exercises, the interflow and interplay between economics and philosophy is strong, whether the contribution is from economists or philosophers. However, the assumptions of the economists are "neoclassical," with their models built on standard rational choice formulation of individual behavior. Nelson (Chapter 6), an economist herself, critiques these assumptions as being inadequate characterizations of individual behavior, making the point that they also restrict policy options available as they do not allow for moral and ethical behavior on the part of individuals. Such behavior is explored further by Lenferna (Chapter 8), who develops the position that there are strong consequentialist and non-consequentialist arguments for divestment from fossil fuels. Karnein (Chapter 7) addresses these issues at the national level and suggests that when a country such as the US refuses to comply with the Paris Agreement, it faces other moral duties as a noncomplier.

As might be expected, the time dimension, looking across generations, also leads to significant overlaps between economists and philosophers, and it also leads to significant debates—within economics and within philosophy as well as between the two. The philosopher Caney (Chapter 9) explores the implications of the normative principle that policy in any generation should be so directed as to leave future populations at least as well off as the present generation. As he notes, this is related to economics Nobel laureate Robert Solow's dictum that sustainability requires each generation to leave to the next "whatever it takes to achieve a standard of living at least as good as our own" (Solow, 1992, p. 15). It differs from Llavador, Roemer, and Silvestre's (2015) perspective, which adds a commitment to improve well-being over time.

The issue of discounting is not a focus in the Caney perspective but is central to many debates in economics and philosophy. Famously, the *Stern Review* (2007) was criticized, and praised, for using a very low discount rate to evaluate the time paths generated by its IAMs. Weitzman (2007) criticized the choice because actual interest rates in the marketplace were much higher, which in turn raised the question of why market interest rates should be used for normative evaluation. Rendall (Chapter 10) takes up the standard reason given in economics as to why there should be discounting of the future—that without it there could be no ranking of alternative time paths extending to infinity; discounting is needed to give finite values so alternatives can be compared. Rendall's answer is to supplement discounting with avoiding the risk of extinction or permanent impoverishment. A version of a similar perspective, centered on his Controllability Precautionary Principle (CPP), is advanced by Pissarskoi (Chapter 11). He argues for interpreting the CPP to require reducing CO_2 concentrations to the pre-industrial level of 280 parts per million, to most decisively avoid devastation of the planet.

Supplementation of a discounted sum of benefits into the future with other morally relevant considerations is also present, as an exposition and a critique, in the focus by Kelleher (Chapter 12) on the standard economic formulation of evaluating time paths of well-being as put forward by Ramsey (1928) which is simply the sum of discounted per capita utilities of each generation multiplied by the number of people in that generation. This is clearly a utilitarian social welfare function, but Kelleher further characterizes the Ramsey approach, which is basically the economist approach as exemplified by Stern (2007), as being one where

> the *betterness* ordering... is *ipso facto* a *rightness* ordering, where a rightness ordering ranks feasible paths in terms of *everything* that matters to policy choice.... Many philosophers will be comfortable with (and all philosophers will be familiar with) the idea... that value considerations cannot *by themselves* fully determine proper choice from among feasible policy options. Climate economists, in my experience, find this a harder idea to make sense of. (Kelleher, Chapter 12)

One example of concern with broader considerations than just the valuation of time paths would be a focus on the injustice of harm done to humans who will be killed, even in the very distant future, as a result of choices about climate change made now. This thesis is developed by Nolt (Chapter 13).

Thus the overlaps and interactions between economic and philosophical perspectives are manifold. On the spectrum from positive to normative analysis of policy options, economics dominates at the positive end while philosophical contributions are primarily at the normative end, especially in highlighting the thinking through of a range of principles for climate justice. Economics relies on these, as seen in the assessment of the equity of the Paris Agreement. But economists appear to be most comfortable with a utilitarian framework of analysis, while philosophers push further and harder for consideration of non-consequentialist frames of evaluation. Further routes to integration will be discussed in Section 1.4, but we now turn to a more detailed account of the chapters in the volume.

1.3. The Volume Chapters

With the narrative framework of Section 1.2, this section provides a brief account of each of the chapters. The volume begins with a chapter that in its very title evokes a concern with the distributional impact of climate change, going against the standard characterization of economists as being concerned only with gross domestic product (GDP) and not with its distribution. Chapter 2, "Poor People on the Front Line: The Impacts of Climate Change on Poverty in 2030" by Julie Rozenberg and Stéphane Hallegatte, develops alternative scenarios with no climate change as the baseline. The drivers of poverty are modeled and applied to data for ninety-two countries. These drivers include demography, sectoral shifts in production, productivity, and economic growth, as well as distributional policy. In the framework of this model, two scenarios, "prosperity" and "poverty," are compared to each other.

Into this set-up they then introduce the impacts of climate change on agricultural incomes, disease, natural disasters, etc. For example, food prices and agricultural incomes are derived from the global agricultural model of Havlík et al. (2015). For malaria, they use the work of Caminade et al. (2014), which provides estimates of malaria increase due to climate change. Given the uncertainty surrounding the quantitative estimates of these impacts, they consider a high-impact and low-impact scenario. This gives them four cases to compare, crossing "prosperity" and "poverty" with high and low climate change impact.

In the poverty scenario, the one where distributional policy is not supportive of poverty reduction, they estimate that 122 million more people will be in income poverty in 2030 with a high impact of climate change, compared with no climate change. At the other end, in the prosperity scenario when distributional policy is supportive, with low impact of climate change there are three million more poor people in poverty at the end of the scenario period in 2030. The authors emphasize that this range primarily reflects policy choices—how the impact of climate change is minimized and how distributional policies work to support poverty reduction. But even these numbers, although they are disaggregated in terms of distribution, are global aggregates. The authors show that the impacts of climate change on poverty are particularly severe in Africa and in South Asia, where levels of poverty are already high.

Chapter 2 plays through the fifteen-year consequences of the climate change that is already "baked into" the system because of past greenhouse emissions. The next two chapters in the volume consider the immediate equity implications of different forms of emissions reductions themselves, without going into the implications of these reductions for climate change in the future. Michael Jakob, Ottmar Edenhofer, Ulrike Kornek, Dominic Lenzi, and Jan Minx address the distributional implications of mitigation in Chapter 3, "Governing the Commons to Promote Global Justice: Climate Change Mitigation and Rent Taxation." The basic idea is that since reducing emissions in effect restricts the use of the global commons of the atmosphere, it creates rents. The question then is who gets the rents and how are they used?

Thus, as Jakob et al. detail, the Paris Agreement to limit temperature increase to below 2°C means a global carbon budget of around 40 gigatons of CO_2 per year. This would require a carbon price in the range of $40 to $80 per ton. Such carbon pricing could generate revenues of the order of $1 trillion to $10 trillion per year, depending on the specific integrated model used, by 2040. This is a significant sum of money and could be used to address a range of distributional issues, including the negative consequences of carbon pricing for the poor of today. Jakob et al. then illustrate the close overlap between economics and philosophy as they take the reader through the basic principles which could guide such revenue recycling. They point out that there is considerable debate not only on revenue distribution within a state but redistribution across states: "In contemporary global justice debates, there is deep disagreement about the extent of such redistributive claims. On the one hand, cosmopolitans argue that robust principles of distributive justice ought to apply globally, while minimalists argue that stronger principles apply within states." They go on to discuss the political feasibility and implementation issues of the taxation of climate and natural resource rents.

Chapter 4, by Adam Rose, Dan Wei, and Antonio Bento, is in a similar spirit to Chapter 3 by Jakob et al. but is more closely related to the Paris Agreement. "Equity Implications of the COP21 Intended Nationally Determined Contributions to Reduce Greenhouse Gas Emissions" looks in detail at the specific proposals in Paris for allocating emission reductions across countries. They first look at the country-specific reduction pledges and compare these to a range of equity principles. Second, they allow trading of emissions allowances across countries. Using a model of trading equilibrium they calculate the price of the emissions allowance and the actual amount of sales and purchases that would occur if trading were allowed. This permits them to once again assess the equity implications of the initial allocation.

After reviewing a range of equity principles, Rose, Wei, and Bento focus on (i) the egalitarian principle and (ii) the ability to pay principle. They interpret the first as saying that emission allowances should be allocated in proportion to population. On the other hand, they interpret the second as saying that emissions reductions should increase with per capita GDP. These, and other more detailed questions, are addressed using the COP21 Intended Nationally Determined Contributions (INDCs) to reduce emissions. These were agreed to by 195 countries, but translation into precise quantitative magnitudes was possible for only a subset of the countries. They then conduct a series of exercises such as comparing the Lorenz curve of the actual INDCs with the implied Lorenz curve of the egalitarian principle and the ability to pay principle. They find that the actual INDCs represent significant departures from these principles as measured by a number of inequality indices. For emissions trading, the model is boiled down to a representation of trading between regions—specifically, data are used for China, the EU, and California. Once again the authors find that equity principles are violated by the Paris INDCs. Indeed, although emissions trading is well understood to reduce global costs of emissions reduction and is thus an efficient policy to pursue, their calculations show that "wealthier countries stand to gain a much higher proportion of the savings, thereby further worsening the equity outcome."

The economist authors of Chapters 2, 3, and 4 build specific models to predict the outcomes of various policies. Their normative evaluative focus varies between individuals within a country (Chapters 2 and 3), across countries (Chapters 3 and 4) and all individuals in the world (Chapter 2). The philosophers Nicole Hassoun and Anders Herlitz bring together these considerations in Chapter 5, "Climate Change and Inequity: How to Think about Inequities in Different Dimensions." As is made clear by the economic models of Chapters 2, 3, and 4, "Climate change and climate negotiations actualize equity considerations in at least three relevant dimensions: distributions of benefits and burdens across countries, within countries and across individuals in the world."

The authors give a number of examples where equity within nations, equity between nations, and equity across all individuals in the world could move in different directions. This can clearly happen in practice for a range of climate mitigation efforts. They then proceed to discuss how the different types of inequality could be weighed relative to each other. Although they do not themselves present arguments as to why the weights might go one way or another, they clearly raise an important set of issues for philosophers to consider. Significantly, there is an overlap with economists' work on decomposing world inequality into a within-countries and between-countries component (see for example Lakner and Milanovic, 2016), and their reflections on inter-country differences as representing inequality of opportunity (Milanovic, 2016).

The economic models of Chapters 2, 3, and 4 have as their base the standard "ECON 101" rational choice model of the individual. Thus the emissions trading mode of Chapter 4, although applied to countries, treats them as individuals in a market where price adjusts to match supply of and demand for emissions allowances. Implicit in the agricultural income model of Chapter 2 is farmers' rational response to climate change, determining prices and incomes in agricultural markets. And although the framework of Chapter 2 admits of externalities and market failures in the allocation of the global commons that is the atmosphere, the overall economic analysis is still in the framework of conventional welfare economics with rational choice individuals. It is these assumptions which are questioned by Julie A. Nelson, an economist, in Chapter 6, "Climate Change and Economic Self-Interest."

Nelson provides a critique of economic approaches from the inside, so to speak, though her critique is focused more on the basic model of choice that is the bread and butter of economics, whether this model is applied to individuals or to nations modeled as individuals. She starts with the position of Christiana Figueres, executive secretary of the UN Framework Convention on Climate Change at the time of Paris, that national economic self-interest would drive nations to agreement. While Nelson concedes that "there is some truth to her argument," she poses the following questions: "Is her assertion about national, and human, behavior true? Is economic self-interest truly the *most powerful* motivator? Is it really the *best* thing to rely on, in our search for a sane and equitable solution to the climate change crisis?" Nelson's answers to these questions pose a challenge to the analytic basis of the economic method.

When an agreement leads to aggregate gains overall but there are gainers and losers under that aggregate, what keeps the losers in the agreement? Economists argue that "side payments" from the winners to the losers would keep the agreement, but as economists who work on "mechanism design" know, even with this there will be incentives to cheat after the agreement is signed. Nelson quotes a well-known book by Posner and Weisbach (2010, p. 170), who are ultimately forced to call on ethics as *deus ex machina*, outside their "incentive compatibility" framework: "But the obligation to achieve a broad, deep, and enforceable treaty imposes a serious ethical duty on rich and poor nations alike—the obligation to cooperate. In our view, it is unethical for a nation to refuse to join a climate treaty in order to free-ride off of others."

Nelson goes on to elaborate on the accumulating evidence in psychology and in behavioral economics, that human beings do indeed have strong ethical motivations. Nelson refers to Sen's (1977, p. 326) famous distinction between "sympathy" and "commitment" as an early warning that all was not well with economic orthodoxy: "If the knowledge of torture of others makes you sick, it is a case of sympathy; if it does not make you feel personally worse off, but you think it is wrong and you are ready to do something to stop it, it is a case of commitment."

Her critique is directed at economists who "seem to be belatedly 'discovering' the social nature of human beings." The award of a Nobel prize in economics to Richard Thaler (2017) holds out the hope that these perspectives may become mainstreamed in economics, including in the economic analysis of climate change. But in many ways Nelson's critique is also hopeful for climate agreements. She concludes by quoting Christiana Figueres to show that her own motivations were not from self-interest, but from deep ethical roots.

The moral basis for action is also the focus of the next two chapters in the volume. In Chapter 7, "Noncompliers' Duties," Anja Karnein explores the possibility that those who have failed to fulfill their basic responsibilities toward climate change, which includes scores of governments and large numbers of individuals, ought to fulfill some specific other responsibilities as a consequence of their noncompliance. Since so many are noncompliers, this is a highly significant issue that is virtually untouched so far by either philosophers or economists. It is, however, especially difficult to make sense of particular duties for noncompliers when the option of compliance with the original duty still remains open. Why wouldn't the whole duty of a noncomplier simply be to become a complier?

Karnein considers several conceptual avenues, including what she labels the complementarity view. The original duty, for example, to reduce emissions a specific amount, should be understood as complemented from the beginning by a duty to, say, show respect for potential victims of climate change by whatever one does. If one complies with the original responsibility, one ought to do so in a respectful manner. And if one chooses not to comply fully with the original duty to reduce emissions to a particular degree, the complementary duty to respect potential victims remains. And then fulfillment of the complementary duty takes on an even more significant signaling role as a means of showing that one's decision not to fulfill the original responsibility is not meant to indicate disregard for those who will be negatively affected by one's failure. If one complies with neither the original duty to reduce emissions nor the complementary duty to show respect for potential victims whatever else one does, one can reasonably be seen as displaying disregard for anyone other than oneself.

In Chapter 8, "Divest–Invest: A Moral Case for Fossil Fuel Divestment," Alex Lenferna emphasizes both the variety of arguments in favor of divestment and the extent to which they complement or overlap with each other. He builds from an influential but philosophically inadequate argument in the activist literature, refining and deepening it, while also mobilizing other types of moral considerations. His fundamental argument is that a moral case can be made for divestment from consequentialist as well as non-consequentialist foundations:

> The consequentialist moral case is grounded both in promotional duties to act on climate change, and the responsibility to not contribute to the potential unnecessary, grave, and substantial harms associated with the fossil fuel industry's continued business model. The non-consequentialist moral case asks institutions to act with integrity and avoid the moral tarnish that comes from investing in, and thus being supportive of and complicit in, the injustices and grave harms entailed in the fossil fuel industry business model.

With regard to the objection that divestment is unlikely to have much effect, Lenferna argues both that one can have a responsibility to act even when the direct effects are imperceptible, and also that major institutions like prominent churches, universities, and pension funds are in a position to exert highly visible leadership that may inspire broader collective action. But he also highlights the argument that divesting may be not only the right thing to do for institutions, but also the smart thing to do since non-fossil forms of energy are an essential part of a prosperous low-carbon future.

In all of the expositions so far, the issue of future generations has been present, but it has not been considered in detail and in depth. For example, whether and, if so, how to discount the future, has been in the background in normative discussion. Even when, for example in Chapter 2 or Chapter 3, time paths of economies were modeled, evaluation of the overall time path, weighting up the present and the future, was not a central focus. But, of course, evaluating the outcomes across generations is indeed central to the economic and philosophical discourse on climate change, and the next block of chapters tackles this issue in its various manifestations.

In Chapter 9, "Justice and Posterity," working from the basic idea that we ought not to leave future people worse off than we are, Simon Caney compares a number of more concrete and detailed interpretations of that fundamental thought. He then explains and defends his own interpretation of a Principle of Justice for Future People as requiring the maintenance of a system of equal capabilities at the highest standard of living which can be enjoyed by those alive at one time and also leave future people with an at least equally good set of capabilities. Unlike Llavador, Roemer, and Silvestre (2015), for example, Caney allows, but does not require, making future people better off. That the core requirement of justice is a maximal equal standard of living is assumed from his own earlier work. But the period of time across which one has responsibility for justice remains to be specified. To do this Caney introduces a causal impact principle that explains why transtemporal duties extend as far as foreseeable effects extend, which in the case of climate change is a matter of centuries. He also explains why the effects should be calibrated on the basis of capabilities and the equality of their distribution. Caney's Principle of Justice for Future People has, then, an absolute component: a duty to promote the absolute standard of living of future people as far as one can reasonably be expected to foresee so that they enjoy a standard of living that is at least as good as we are entitled to have. And the principle also has a distributive component: a duty to promote the realization of equality *within* future generations.

The issue of discounting is taken on full force by Matthew Rendall in Chapter 10, "Discounting and the Paradox of the Indefinitely Postponed Splurge." One of those things that "everybody knows to be true" is that unless we discount future benefits, we would never enjoy any benefit now, because we would invest, not consume. Any attempt to forego a positive rate of time preference leads, according to Tjalling Koopmans, to the "paradox of the indefinitely postponed splurge" (PIPS). Except that, as Matthew Rendall argues convincingly in his internal critique of Koopmans' argument, it doesn't. Rendall shows that the argument for the PIPS "falls apart," establishing that undiscounted utilitarianism is not a self-defeating theory.

But Rendall not only dismantles the argument for discounting based on PIPS, he also offers penetrating critiques of discounting as a utilitarian decision-making procedure. Against the argument that one ought to discount at the market interest rate lest one make inefficient investments that leave future generations worse off, Rendall notes that Martin Weitzman's defense explicitly assumes that the real interest rate remains the same across generations. But one cannot assume that the real interest rate will remain the same when a catastrophe caused by climate change could cause long-term growth to stall. And because climate change depends on stocks as well as flows, one also cannot assume that one can switch it off once catastrophe appears likely. This inattention to possible disaster is one reason why "discounting as usually practiced, then, does not maximize long-run expected utility." Put positively, Rendall's thesis is that "it is far more important to minimize the risk of such catastrophes than to increase consumption." This leads him to endorse a constraint similar to Hans Jonas's imperative of responsibility, "which forbids us from making 'the existence or the essence of man as a whole... a stake in the hazards of action'."

Like Rendall, Eugen Pissarskoi argues for focusing climate policy on the avoidance of catastrophe. In Chapter 11, "The Controllability Precautionary Principle: Justification of a Climate Policy Goal Under Uncertainty," he presents challenges to the arguments of both many economists and many philosophers. Pissarskoi puts aside the subjective probabilities regularly employed by economists and takes seriously the fact that the objective probability for any specific value for climate sensitivity (the extent to which climate will respond to increases in forcing by accumulated greenhouse gases) is unknown, leaving open the possibility that climate sensitivity is much higher than conventionally assumed. And while he is sympathetic to the possibilistic approach using a precautionary principle invoked by a number of philosophers, he shows that their versions of the precautionary principle fail to justify choosing to allow the atmospheric concentration of CO_2 to stabilize at as high a level as 550 ppm because, if climate sensitivity is actually higher than usually assumed, a concentration of 550 ppm risks catastrophe just as higher (and some lower) concentrations do.

In response Pissarskoi formulates a new precautionary principle that introduces conditions on the process by which a policy allows catastrophe to be risked. Policies that allow higher concentrations of CO_2 allow the risk of catastrophe as the result of uncontrollable natural climate responses. Policies that drive concentrations of CO_2 lower by means of negative emissions risk catastrophe through disruption of social and economic processes that humans may nevertheless be able to control. He suggests that the latter risks of catastrophe from potentially controllable social processes may be less dangerous than the risks of catastrophe from uncontrollable natural processes.

J. Paul Kelleher uses the concept of social cost of carbon, and the associated issue of discounting the well-being of future generations, to open up the question of the difference between choice and valuation, the latter being much broader than simply a discounted sum of per capita consumptions of all generations. In Chapter 12, "The Social Cost of Carbon from Theory to Trump," Kelleher emphasizes that the social cost of carbon (SCC) is calculated using a standard (Ramseyan) kind of value function (or social welfare function) that assumes that a ranking of betterness based on comparing sum totals of utility from consumption just is a rightness ordering for policy, which amounts to the highly controversial assumption that nothing else should matter for policy except utility from consumption. (This utilitarian assumption is rejected by many contributors to this volume, including Kelleher.) "*V-functions merely place consumption paths into an ordering*. What that ordering signifies morally, and what public policy is to do with that ordering, are distinct and further questions." Since "the concept of the social cost of carbon is yoked to the idea of a value function... a gap between the policies that V ranks highest and the policies that we should pursue all-things-considered" means that "the SCC will not be an ironclad guide to policy choice." Indeed, "since the SCC depends so heavily on the *V*-function, it could be a giant mistake—both analytically and dialectically—to speak of 'the' SCC... Only an SCC tied to a *V*-function capturing all relevant normative considerations should be used to identify policies that are 'optimal' in the sense relevant to final social choice."

In the course of his argument Kelleher distinguishes a "Ramsey rule" that is committed to a utilitarian value theory and a "Ramsey rule" that is not. He then goes on to distinguish "three families of value functions—generalized utilitarian, prioritarian, and maximin," all of which are ethical observer, or social planner, approaches, from representative agent models. Next he examines the significance of the contrast between descriptivist (revealed preference) and prescriptivist calibrations of value parameters, outlining major weaknesses in Weitzman's descriptivist critique of Cline and Stern and providing his own methodological critique of the *Stern Review*. He concludes with comparative assessments of the work on the SCC by the Interagency Working Group (IWG) during the Obama Administration (with a final SCC of $51), the examination of the IWG's work that it requested from the National Academy of Sciences, and OMB Circular A-4 from the Bush Administration that the Trump Administration substituted for the IWG's SCC, which includes a 7 percent discount rate based on the opportunity cost of capital and counts only domestic climate damage, and so yields an SCC of $1.

The volume concludes with Chapter 13, "Long-term Climate Justice." Challenging many economists and many fellow philosophers alike, John Nolt argues that, in the case of climate change, intergenerational justice is not best understood as primarily a matter of fair or unfair distribution. Climate change "will displace, sicken, injure and kill huge numbers of people over the coming centuries, perhaps millennia. This is an unprecedented injustice," but this injustice is not mainly a matter of unfair distribution. The injustices being produced by climate change are straightforward harms, including fatal harms, that theories of distribution do not adequately capture.[2] An alternative to distributional accounts of climate justice is climate justice as respect for human rights, which is appealing insofar as it focuses on prohibiting harm. But a strict prohibition on harm is unfulfillable, and in the case of rights theory applied to future generations, the unfulfillable rights would proliferate.

A more plausible theory is what some philosophers call a "consequentialism of rights" that focuses on objective welfare and does not attempt to maximize anything, but aims to "satisfice": "eliminating the worst options and choosing, from among the better, one that is good enough." The consequentialist aspect of this rights theory brings it somewhat closer to the ordinary consequentialism espoused by many economists, but unlike that ordinary consequentialism, the consequentialism of rights does not permit serious harms to be outweighed by their benefits. A rights-violating harm can be justified only when necessary to the avoidance of a more serious rights-violating harm. Like Nelson, Nolt believes it is crucial to specify accounts of justice that can motivate, and he maintains that "we can be motivated by our natural revulsion at the spectacle of the powerful, heedless, and arrogant unjustly harming people who are powerless, voiceless, and without recourse against them—even when the powerful, heedless, and arrogant are us."

1.4. Pathways Toward Integration

Obviously, then, the contributors to this volume do not line up, as in a football match, as "the Economists vs. the Philosophers." Quite the contrary, they work more as a team, although, as usual, with different individuals with different talents playing different roles in the formulation of responses to climate change. Their integration occurs at different levels and takes different forms and degrees—there is no one formula for integrating economic and philosophic considerations. Some examples from the volume will suffice to illustrate the interplay and interaction that is already happening. We conclude with a major issue on which a joint effort by economists and philosophers is necessary and possible—the sustainability of the Paris Agreement.

In showing that alternative development strategies play a crucial role in channeling the effects of climate change on those in poverty, Rozenberg and Hallegatte provide valuable concrete empirical guidance needed by those who, like Hassoun and Herlitz, argue for the application of abstract normative standards of equity, interpreted as various types of distribution, to the outcomes

[2] For other arguments that business as usual and failure to ratchet up mitigation urgently constitute the infliction of harm on future generations, see Shue 2017 and Shue 2018.

of climate policy. Every plausible principle of justice gives some degree of priority to the needs of the poor. While Rozenberg and Hallegatte concentrate on inequality within individual countries (while using purchasing power parity (PPP) exchange rates for international comparison), Hassoun and Herlitz expand the focus to include additional facets of equity, and especially inequality, that economists do also take into account.

Caney tackles the daunting task of specifying exactly what individuals who live in an earlier generation owe to individuals who live in later generations. In order to do this he specifies the "currency" of what is normally labeled "intergenerational justice" (although Caney maintains that both those with the responsibilities and those with the entitlements are individuals, not literally generations). He suggests that the appropriate "currency" is capabilities, thereby adopting the capabilities approach that was formulated jointly by one economist (Amartya Sen) and one philosopher (Martha Nussbaum). This directly challenges those other economists who wish to evaluate the effects of climate change "solely for their impact on GDP or resources." And because his answer to the question "what is owed to future people?" is simply "an at least equally good set of capabilities" as the set of capabilities of current people, he offers an alternative to economists who contend that current people ought to leave future people better off.

Rose, Wei, and Bento examine the emissions reductions pledged in the bottom-up process adopted for Paris 2015, while Jakob et al. examine the revenues that could be collected from a mechanism like a carbon tax to bring about emissions reductions. In both cases, however, the focus is sharply on distribution, equity, and fairness. Noting that national positions in climate negotiations "have in fact revolved around equity explicitly and implicitly," Rose, Wei, and Bento assess both the initial allocation of effort consisting of the actual pledges and the likely outcomes of trading starting from those pledges. Their assessments employ both equity metrics common within economics (the Gini coefficient and Atkinson index) and equity principles common in philosophical positions, especially an egalitarian principle and a version of ability to pay. This is one form of thorough integration of economic and philosophical principles. This enables them to show not only that in Paris 2015 the poorer countries acquiesced in greater departures by the initial pledges from the equity principles they believe in—for example, industrialized countries did not come close to pledging emissions reductions in accord with ability to pay—but also that the process of trading emissions permits on the basis of these bottom-up pledges would worsen international inequality to the further disadvantage of the poorer countries (while reducing the total aggregate costs). "The countries/regions with the highest current GDPs reap the greatest rewards of trading," producing outcomes directly counter to, for example, the requirements of equity understood as contribution according to ability to pay. Such increasing inequities certainly might cause the Paris Agreement to unravel.

Jakob et al., like Lenferna, assess a particular strategy for tackling climate change. The former formulate and evaluate the introduction of rents on the use of the atmosphere as a disposal dump for greenhouse gases, in the form of carbon taxes for the sake of reducing emissions, while distributing the funds collected in a manner that promotes social justice by financing access to infrastructure for the poorest members of society. Thus they too thoroughly integrate the mitigation of climate change with the promotion of sustainable development, like Rozenberg and Hallegatte, at the same time that they thoroughly integrate economic considerations with social justice. If Rose, Wei, and Bento are correct that unadjusted trading would increase inequality between industrialized countries and poorer countries, the distributions urged by Jakob et al. would be one means of reducing international inequalities.

Just as Rose, Wei, and Bento, who are economists, structure their research around normative principles formulated by philosophers and political theorists, Rendall and Kelleher, who are not economists, critically examine concepts central to economics. Both Rendall and Kelleher make normative arguments, but the subjects of their respective arguments, discounting and the SCC, are at the heart of economists' conventional approaches to climate change. Rendall and Kelleher each fundamentally challenge aspects of these concepts as economists normally use them. Nelson, who is an economist, also challenges a central assumption of the standard economic approach to climate change.

Differing from many of the other authors here, Nolt challenges economists and philosophers alike to consider to what extent the effects of climate change are better captured through a lens of harm rather than a lens of distribution. He is not suggesting that distributive issues should be ignored, but he is contending that legitimate concerns with questions of fairness should not be allowed to obscure the primary fact that climate change is harming millions of people and the generally shared principle that the infliction of such harm is straightforwardly wrong.

Thus it would appear that economists and philosophers are already beginning to play as a team in addressing the issue of climate justice. However, there is one central issue around which it will be necessary for them to come together to provide the conceptual and empirical basis for a policy accord, namely, sustaining the Paris Agreement on voluntary emissions reductions. This requires current generations to sacrifice for the benefit of the future. Several chapters in this volume have addressed this question, directly or indirectly. Could economics and philosophy forge a joint agenda of analysis and policy dialogue based on a common platform? We think this is possible.

In 1975, economist Robert L. Heilbroner wrote a widely discussed piece for the *New York Times* titled "What Has Posterity Ever Done for Me?" He began:

"Will mankind survive? Who knows? The question I want to put is more searching: Who cares?" And he concluded: "Yet I am hopeful that in the end a survivalist ethic will come to the fore... from an experience that will bring home to us... the personal responsibility that defies all the homicidal promptings of reasonable calculation."[3] "Who cares?"—or perhaps better, "why would we care?"—is a question about motivation, and Heilbroner's ultimately hopeful answer appeals to a sense of personal responsibility. What could this mean now in the context of climate change?

In her chapter here Nelson has argued, correctly we think, that as a matter of fact many ordinary people are actually moved by considerations other than rational self-interest. An economist who assumes otherwise is making a simplistic mistake about how individuals and societies work. And a philosopher who merely lays out principles without concerning herself with what might lead anyone to abide by those principles is, as John Nolt's chapter observes, leaving her argument unfinished and her case not fully made. Several others here have appealed to a sense of responsibility of one kind or another.[4] One of the challenges that future work on climate justice by both economists and philosophers needs to take up is the much deeper exploration of what, if anything, can motivate people alive now—individual persons and governments—to feel responsibility and to take action on behalf of people who will live much later. Here we can offer, not a road map, but simply the slightest illustrative hints at possible paths.

The fundamental problem, crudely put, is that much—not all, but significant portions—of what needs to be done by people who live in the present, to protect people who will live in the future against the threats created by climate change, will prevent harms that will otherwise affect only the people in the future and will impose only costs on the people who act now. One element that seems to be missing is any hope of reciprocity. Hence, Heilbroner's question: what has posterity ever done for me? The little joke embedded in Heilbroner's question relies on a tacit assumption about the importance of reciprocity: that it would not be reasonable for me to do something for people who cannot possibly do anything for me.

One can challenge the whole assumption that some kind of reciprocity is necessary to underwrite justice in general and climate justice in particular, but it seems more worthwhile to explore first which kinds of reciprocity are available and whether they could ground motivation for just action toward individuals whom we can affect profoundly but who cannot affect us. Philosopher Stephen Gardiner has wrestled with what he calls "the tyranny of the contemporary" and has observed that if one treats the relation between

[3] Heilbroner (1975, pp. 169 and 175–6).

[4] Anja Karnein in this volume extends the discussion to duties of non-compliers.

various generations' choices about emissions reductions as an intertemporal prisoner's dilemma—"it is *collectively rational* for most generations to cooperate [by reducing emissions]," but "it is *individually rational* for all generations not to cooperate"—"the standard solutions to the prisoner's dilemma are unavailable: one cannot appeal to a wider context of mutually-beneficial interaction, nor to the usual notions of reciprocity."[5] One possibility is that the conceptual poverty of simpler notions of reciprocity is part of the problem.

The crudest version of the usual notion of reciprocity is that it is a positive form of tit-for-tat: if you scratch my back, I will scratch yours. But this vulgar understanding is little different from a market exchange between a willing buyer and a willing seller: you provide me a meal, and I will provide you the price of the meal. While reciprocity does involve some form of conditionality, it need be neither nearly so blatant and direct nor bilateral. It may be that after a couple of years in which I do not send you a birthday card, you may stop sending me a birthday card. But we would not make an explicit "deal" upfront: I would be willing to send you cards, but only if you are going to send me cards. Reciprocity does involve "exchanges" of certain kinds, but often they are exchanges of gifts. Part of what makes them "gifts," not purchases paid for in-kind, is that there is no "deal"—no explicit and direct conditionality. And they by no means need to be bilateral. Under one of the most sophisticated and complete conceptions of reciprocity, reciprocity is a "retrospective obligation" in which good received from predecessors can be "returned" as good provided to posterity, thus involving a minimum of three generations (Becker, 1990).

A few years earlier than Heilbroner's article, Richard Titmuss's classic study of blood donation in the British National Health Service (NHS), *The Gift Relationship*, reflected the fact that while many people who donate blood take for granted that if they themselves need blood, they will receive it, the connections between givers and receivers are far looser than any tit-for-tat picture, and they are not bilateral.[6] It is not the case, for example, that one can receive four pints of blood during surgery only if one has previously donated at least four pints, or even only if someone—oneself or others on one's behalf—will later donate at least four pints to "pay back," or even only if one ever donates any blood at all. On the contrary, one receives the blood one needs as long as it is available, and "blood drives" are held precisely in the hope of having sufficient stocks available for emergencies. On the other hand, often afterwards a grateful recipient who had not previously been a donor will choose to become a donor in order to "pay back" the system. Many people neither give nor receive, while some faithful donors donate many times over

[5] Gardiner (2016, pp. 9, 27, and 28).
[6] Titmuss (1970, 1997) Also see Arrow (1972) and Singer (1973).

the years, but never receive any blood in return. And so on. This might be called loose-jointed, multilateral reciprocity.

Titmuss had emphasized his conviction that societies ought to be structured so that individuals are "free to choose to give to unnamed strangers."[7] In a thoughtful review article on *The Gift Relationship*, economist Kenneth Arrow labels this "impersonal giving" and refers to what Titmuss is advocating as "impersonal altruism."[8] These labels make clear that the NHS blood donation system as described by Titmuss is not at all a matter of bilateral tit-for-tat. But Arrow has from the start located his "impersonal altruism" within a broader category of "unilateral transactions": "the donation of blood for transfusions is only one example of a large class of unilateral transactions in which there is no element of payment in any direct or ordinary sense of the term."[9] Treating blood donation as unilateral seems to miss the point, however, and thinking of Titmuss's gifts to unnamed strangers as unilateral begs the question against any interpretation of the practice of blood donation as a complex form of reciprocity. If it is unilateral, it is not reciprocal. But why would it seem unilateral? Apparently, only because it is not tit-for-tat (and is in that very specific sense, "impersonal"). We have no reason at all, however, to operate with a simplistic dichotomy consisting only of unilateral and tit-for-tat.

In elaborating his own conception of reciprocity, philosopher Lawrence Becker speaks, not of "unilateral" giving, but of "the indefiniteness of our intention in giving blood."[10] Among the many things that are neither unilateral nor tit-for-tat is what international lawyer Mark Osiel, writing about how international laws against torture can work, calls "diffuse or systemic" reciprocity, which brings "gains from forbearance [from torture] that emerge at the multilateral level over a longer term."[11]

Now, why some individuals continue to donate blood when some others do not donate, and why some sovereign states forbear from using torture when some other sovereigns do not forbear from inflicting torture—and how much non-cooperation it takes to undermine the system—are each stories too long to be told here. What we have gained already, however, is a richer and more sophisticated conception of reciprocity that might help us to understand the Paris Agreement of 2015 and what might motivate nations to comply with it.

After a quarter century of futile effort to arrive at a multilateral climate treaty with binding emissions quotas, the COP surrendered to unrelenting US pressure and switched from a top-down to a bottom-up treaty, as explained in this volume by Rose, Wei, and Bento, issuing in the INDCs. As they show, the INDCs—they are now Nationally Determined Contributions (NDCs)—so far do not satisfy any of the most widely held normative standards for shared

[7] Titmuss (1970, p. 242; 1997, p. 310). [8] Arrow (1972, pp. 346 and 360).
[9] Arrow (1972, p. 345). [10] Becker (1990, p. 231). [11] Osiel (2009, p. 368).

national responsibility, and of course at the time of writing the US federal government is threatening to back out of its own chosen NDC on the grounds that keeping it is not in its own interest. If the US defects from the Paris Agreement, as it currently says it will, what could motivate other parties not to defect as well?

Time will tell, of course. But the most heartening fact so far is that the response to the US federal government's announcement of its intended defection has not only not been widespread defections by other governments, but affirmations and reaffirmations by many others of their intention to stay the course as well as widespread efforts by corporations and by US governments at all levels except the federal level to step into the breach created by Washington.[12] It is early days, and much remains to be seen. Our only suggestion is that it might be worth exploring a few points of analogy (amid the multiple points of obvious disanalogy) with diffuse or systemic practices of reciprocity, like practices of voluntary blood donation by individuals and choices by sovereign states not to use torture, to see how the motivation to carry on with the effort to prevent dangerous climate change might be found. A number of angles here seem worth exploring in future research.

Titmuss commented that blood donors "signified their belief in the willingness of other men to act altruistically in the future and to combine together to make a gift freely available."[13] Could this be Heilbroner's "personal responsibility that defies all the homicidal promptings of reasonable calculation"? One possible interpretation of Titmuss's observation is that many people comprehend a functioning practice of diffuse reciprocity that accomplishes a valuable function in society (supplying blood according to need) and understand that it can be kept alive in the future if sufficient numbers of individuals continue to act as many have in the past. The precise motivation, which almost surely varies across individuals, would need to be studied—it might focus on the value of the social practice itself or on the good accomplished for individuals by the social practice. In the philosophical conception of reciprocity constructed by Becker, individuals understand that predecessors intentionally acted in ways designed to benefit them and respond by intentionally acting in ways designed to benefit posterity; "the disposition to reciprocate . . . rejects the logic of freeriders."[14] Becker argues that there is a rational basis for this rejection of free-riding in the general conception of morality. At the level of individual behavior, the recent Nobel Prize in economics awarded to Richard Thaler (2017) shows a revolutionary opening in that discipline to considerations going beyond the conventional

[12] See, for example, Bloomberg and Brown (2017). Also compare the movement calling itself "We Are Still In": <https://www.wearestillin.com/us-action-climate-change-irreversible>.

[13] Titmuss (1970, p. 239; 1997, p. 307).

[14] Becker (1990, pp. 229–51, at p. 243).

model of choice in neoclassical economics, as presaged by Nelson in her chapter for this volume.

Parties can make unconditional pledges of reciprocity and conditional pledges of reciprocity—both are found in the NDCs for the Paris Agreement. Reciprocity can be retrospective (Becker)—they did, so I will—or prospective (Titmuss)—I will because I hope (and believe) they (enough of them) will. The precise psychology of a response of reciprocity could have either more negative or more positive interpretations, or both. To realize that one is a moral dead end, or moral black hole, into which good has flowed from predecessors but out of which nothing good flows toward posterity, might be incompatible with a sense of being a worthwhile person. It might arouse shame or a sense of dishonor, and one might act on the negative motivation of escaping the sense of being morally useless. For some individuals, a life of only receiving and not giving might violate a sense of fair play. One might act to escape a sense of being socially parasitic, or act more positively to do one's fair share in a practice in which many others have done theirs. More positively still, one might realize that one has the opportunity to contribute to the preservation of a valuable practice—or the creation of a new one—and take delight in that prospect. Others clearly have no sense of fair play whatsoever and cannot understand the point of not "always getting the best deal possible." And so on.

Our question is whether anything remotely analogous could function to motivate compliance with—and urgent ratcheting up of—the Paris Agreement. Obviously national leaders who have no sense of shame, decency, or fair play will not care that their policies would make their countries into black hole free-riders who only take and give nothing. But at least the initial response to the threatened defection from the Paris Agreement by the current US federal government on the part of all other levels of government in the US, as well as on the part of other national governments and many major corporations, has been to try to keep alive and nurture the practices defined in the agreement.

Is it possible, and is it likely, that a broad international consensus will develop that the Paris Agreement has begun to define a valuable practice of diffuse reciprocity in the reduction of emissions such that, if most parties do their fair share, people in the present can accomplish good, or at least prevent serious dangers and harms, for people in the future? What does this depend on? Does there need to be a consensus on how to characterize fair shares? Or, as in the case of blood donation, can vigorous displays of commitment by some leading nations, even acting on diverse understandings of standards of fairness, motivate sufficiently strong reciprocal action by others? How does this relate to what game theorists call non-cooperative games? And how could the theoretical discourse be operationalized? These seem to us to be a few of the questions on which economists and philosophers—and others—might cooperate in trying to study in the concrete case of the Paris Agreement.

References

Arrow, Kenneth J. 1972. "Gifts and Exchanges." *Philosophy & Public Affairs*, 1: 343–62.

Becker, Lawrence C. 1990. *Reciprocity*. Chicago: University of Chicago Press.

Bloomberg, Michael R. and Jerry Brown. 2017. "The U.S. is Tackling Global Warming, Even if Trump Isn't." *New York Times*, 14 November, <https://www.nytimes.com/2017/11/14/opinion/global-warming-paris-climate-agreement.html>.

Caminade, Cyril, Sari Kovats, Joacim Rocklov, Adrian M. Tompkins, Andrew P. Morse, Felipe J. Colón-González, Hans Stenlund, Pim Martens, and Simon J. Lloyd. 2014. "Impact of Climate Change on Global Malaria Distribution." *Proceedings of the National Academy of Sciences* 111 (9): 3286–91.

Gardiner, Stephen M. 2016. "In Defense of Climate Ethics." In *Debating Climate Ethics*, edited by Stephen M. Gardiner and David A. Weisbach. New York: Oxford University Press.

Havlík, Petr, Hugo Valin, Mykola Gusti, Nicklas Forsell, Mario Herrero, David Leclère, Nikolay Khabarov, Aline Mosnier, Michael Obersteiner, and Erwin Schmid. 2015. "Climate Change Impacts and Mitigation in the Developing World: Integrated Assessment of Agriculture and Forestry Sectors." Forthcoming as a World Bank Policy Research Working Paper.

Heilbroner, Robert. 1975. *An Inquiry into the Human Prospect, with "Second Thoughts" and "What Has Posterity Ever Done for Me?"* New York: W.W. Norton.

Kanbur, Ravi. 2015. "Education for Climate Justice." The Many Faces of Climate Justice: An Essay Series on the Principles of Climate Justice. Mary Robinson Foundation-Climate Justice. <https://www.mrfcj.org/wp-content/uploads/2015/09/Education-for-Climate-Justice.pdf> accessed 11 March 2018.

Lakner, Christoph and Branko Milanovic. 2016. "Global Income Distribution: From the Fall of the Berlin Wall to the Great Recession." *World Bank Economic Review*, World Bank Group, vol. 30(2), pages 203–32.

Llavador, H., Roemer, J. E., and Silvestre, J. 2015. *Sustainability for a Warming Planet*. Cambridge, MA: Harvard University Press.

Milanovic, Branko. 2016. *Global Inequality: A New Approach for the Age of Globalization*. Cambridge, MA: Belknap Press of Harvard University Press.

Osiel, Mark. 2009. *The End of Reciprocity: Terror, Torture, and the Law of War*. Cambridge: Cambridge University Press.

Posner, Eric A. and David Weisbach. (2010). *Climate Change Justice*. Princeton, NJ: Princeton University Press.

Ramsey, F. P. 1928. "A Mathematical Theory of Saving." *The Economic Journal*, 38 (152): 543–59.

Sen, Amartya. 1977. "Rational Fools: A Critique of the Behavioral Foundations of Economic Theory." *Philosophy and Public Affairs* 6 (Summer): 317–44.

Shue, Henry. 2017. "Climate Dreaming: Negative Emissions, Risk Transfer, and Irreversibility." *Journal of Human Rights and Environment*, 8: 203–16.

Shue, Henry. 2018. "Mitigation Gambles: Uncertainty, Urgency, and the Last Gamble Possible." *Philosophical Transactions of the Royal Society A*, 376: 20170105.

Singer, Peter. 1973. "Altruism and Commerce: A Defense of Titmuss against Arrow." *Philosophy & Public Affairs*, 2, 312–20.

Solow, Robert. 1992. *An Almost Practical Step Toward Sustainability*. Washington, DC: Resources for the Future.

Stern, N. 2007. *The Economics of Climate Change: The Stern Review*. Cambridge: Cambridge University Press.

Thaler, Richard. 2017. "From Cashews to Nudges: The Evolution of Behavioral Economics." Nobel Prize Lecture delivered on 8 December. <https://www.nobelprize.org/nobel_prizes/economic-sciences/laureates/2017/thaler-lecture.html>.

Titmuss, Richard M. 1970. *The Gift Relationship: From Human Blood to Social Policy*. London: George Allen & Unwin.

Titmuss, Richard M. 1997. *The Gift Relationship: From Human Blood to Social Policy*, expanded and updated edition, edited by Ann Oakley and John Ashton. New York: New Press.

Weitzman, M. L. 2007. "The Stern Review of the Economics of Climate Change." *Journal of Economic Literature*, 45 (3), 703–24.

2

Poor People on the Front Line: The Impacts of Climate Change on Poverty in 2030

Julie Rozenberg and Stéphane Hallegatte

2.1. Introduction

Estimates of the economic cost of climate change have attracted interest from policymakers and the public and generated heated debates.[1] These estimates, however, have mostly been framed in terms of the impact on country-level or global GDP, which does not capture the full impact of climate change on people's well-being.

In particular, climate change impacts will be highly heterogeneous within countries. If impacts mostly affect low-income people, welfare consequences will be much larger than if the burden is borne by those with a higher income. Poor people have fewer resources to fall back upon and lower adaptive capacity. And—because their assets and income represent such a small share of national wealth—poor people's losses, even if dramatic, are largely invisible in aggregate economic statistics.

Investigating the impact of climate change requires an approach that focuses on people that play a minor role in aggregate economic figures and are often living within the margins of basic subsistence. Macroeconomic models such as Computable General Equilibrium models alone cannot assess the impact on poverty, and a microeconomic approach that explicitly represents the livelihoods of poor people is required. Fortunately, micro-simulation techniques (Olivieri et al., 2014; Bussolo, De Hoyos, et al., 2008; Bourguignon

[1] In this chapter, extreme poverty is defined using a consumption poverty line at $1.25 per day, using 2005 PPP exchange rates. Using the new poverty line (with 2011 PPP rates and the $1.90 line, see World Bank, and International Monetary Fund, 2014) would not change our results in a quantitative manner, considering the consistency between the two PPP rates (Jolliffe and Prydz, 2015) and the way we treat the uncertainty on poverty numbers.

et al., 2005) and the generalization of household surveys now make it possible to look at this issue. Innovative steps in this direction have been made recently, combining aggregate impact estimates with household-level data or using specific models (Hertel and Rosch, 2010; Hertel et al., 2010; Ahmed et al., 2009; Gupta, 2014; Jacoby et al., 2014; Skoufias et al., 2011; Devarajan et al., 2013; Bussolo, de Hoyos, et al., 2008). Most of these analyses investigate the effect of climate change on poverty through agricultural production.

Assessing the impact of climate change on poverty remains, however, a daunting task. In particular, the speed and direction of future socio-economic changes will determine the future impacts of climate change on poor people and on poverty rates as much as climate change itself (Hallegatte et al., 2011). It is not hard to imagine that, in a world where everyone has access to water and sanitation, the impacts of climate change on water-borne diseases will be smaller than in a world where uncontrolled urbanization has led to widespread underserved settlements located in flood zones. Similarly, in a country whose workers mostly work outside or without air conditioning, the impact of temperature increase on labor productivity will be stronger than in an industrialized economy. And a poorer household will have a larger share of its consumption dedicated to food and will therefore be more vulnerable to climate-related food price fluctuations than a richer household.

Assessing the future impacts of climate change therefore requires an exploration of future socio-economic development pathways, in addition to future changes in climate and environmental conditions.

First, we assume that there is no climate change, and we explore possible counterfactual scenarios of future development and poverty eradication. Second, we introduce climate change into the picture and look at how it changes the prospect for poverty eradication.

It is impossible to forecast future socio-economic development. Past experience suggests we are simply not able to anticipate structural shifts, economic crises, technical breakthroughs, and geopolitical changes (Kalra, Gill, et al., 2014). In this chapter, we neither *predict* future socio-economic change nor the impact of climate change on poverty. Instead, we follow the approach that is the basis of all Intergovernmental Panel on Climate Change (IPCC) reports, namely we analyze a set of *socio-economic scenarios* and we explore how climate change would affect development in *each* of these scenarios. These scenarios do not correspond to particularly likely futures. Instead, they are possible and internally consistent futures, chosen to cover a broad range of possibilities to allow for an exploration of possible climate change impacts. People sometimes refer to these scenarios as "what-if" scenarios, as they can help answer questions such as "*what* would the climate change impact be *if* socio-economic development followed a given trend." The goals are to better understand how the impact of climate change on poverty depends on socio-economic development, to estimate the potential impacts in "bad" scenarios, and to explore possible policy options to minimize the risk that such bad scenarios actually occur.

We start by analyzing the drivers of future poverty, and we explore the range of possible futures regarding these drivers to create hundreds of socio-economic scenarios for each of the ninety-two countries we have in our database. This analysis combines household surveys (from the I2D2 database) and micro-simulation techniques (Olivieri et al., 2014; Bussolo, De Hoyos, et al., 2008; Bourguignon et al., 2005) and is performed in a framework inspired from robust decision-making techniques, in which all uncertain parameters are varied systematically to the full range of possible outcomes. We combine assumptions on future demographic changes (How will fertility change over time? How will education levels change?); structural changes (How fast will developing countries grow their manufacturing sector? How will the economies shift to more services?); productivity and economic growth (How fast will productivity grow in each economic sector?); and policies (What will be the level of pensions? How much redistribution will occur?). The range of possible futures of these parameters is determined based on historical evidence, and on the socio-economic scenarios currently developed for the analysis of climate change, the Shared Socio-Economic Pathways (SSPs) (Kriegler et al., n.d.; O'Neill et al., 2013). These sets of assumptions are used to generate hundreds of scenarios for the future socio-economic development of each of the ninety-two countries.

Then, we select two representative scenarios per country, one optimistic and one pessimistic in terms of poverty reduction and changes in inequality, and we aggregate them into two global scenarios. The first scenario is labeled "prosperity" and represents a world with universal access to basic services, a reduction in inequality, and the reduction of extreme poverty to less than 3 percent of the world population (this is one of the two official goals of the World Bank Group). A second scenario is labeled "poverty" and represents a world where poverty is reduced, but not to an extent consistent with the goals of the international community, where access to basic services improves only marginally and inequality is high.

Finally, we introduce quantified estimates of the impacts of climate change on agriculture incomes and food prices, on natural hazards, and on malaria, diarrhea, and stunting into these two scenarios. With a 2030 horizon, impacts barely depend on emissions between 2015 and 2030 because these affect the magnitude of climate change only over the longer term, beyond 2050. Regardless of socio-economic trends and climate policies, the mean temperature increase between 2015 and 2035 is between 0.5 and 1.2°C—and the magnitude depends on the response of the climate system (IPCC, 2013). The impacts of such a change in climate are highly uncertain and depend on how global climate change translates into local changes, on the ability of ecosystems to adapt, on the responsiveness of physical systems such as glaciers and coastal zones, and on spontaneous adaptation in various sectors (such as adoption of new agricultural practices or improved hygiene habits). We thus consider again two cases, one high impact and one low impact, and investigate the potential impact of unmitigated climate change on the number of poor people living in extreme poverty in 2030 in the four cases (optimistic and pessimistic regarding socio-economic trends, and optimistic and pessimistic regarding climate change impacts).

We find that climate change has a visible impact on our poverty projections for 2030, even if it remains a secondary driver of poverty trends (compared with policies and socio-economic trends). We also find that, by 2030 and in the absence of surprises on climate impacts, inclusive climate-informed development can prevent most (but not all) of the impacts on poverty.

2.2. Step 1: Exploring the Uncertainty on Future Poverty and Inequality

The first stage of our methodology is to build, for each country, many baseline scenarios for possible future socio-economic change, in which we assume that there is no climate change. To do so, we explore a wide range of uncertainties on future structural change, productivity growth, demographic changes, and policies, to create several hundred scenarios for future income growth and income distribution in each country.

We start from a collection of ninety-two household surveys, describing the population of ninety-two countries and harmonized for the year 2012. These surveys include tens of thousands of individuals per country, which are all assigned a weight so that the entire population is modeled. We then use a micro-simulation model to project the pathway of each individual of these household surveys, between 2012 and 2030. We change the income and weight of each household in the model to reflect macroeconomic changes.

Due to the uncertainty inherent in projecting these changes, we work with ranges of values based on historical data and trends, and on previous work on socio-economic scenarios such as the development of the SSPs (details on this method can be found in Rozenberg and Hallegatte, 2015).

Figure 2.1 shows the result of these simulations for Haiti, Malawi, the Philippines, and Turkmenistan: each dot shows the position of one scenario in 2030 for the selected country, with the horizontal axis representing the average income of the bottom 20 percent in 2030 (as an indicator for poverty) and the vertical axis representing the difference between the income growth

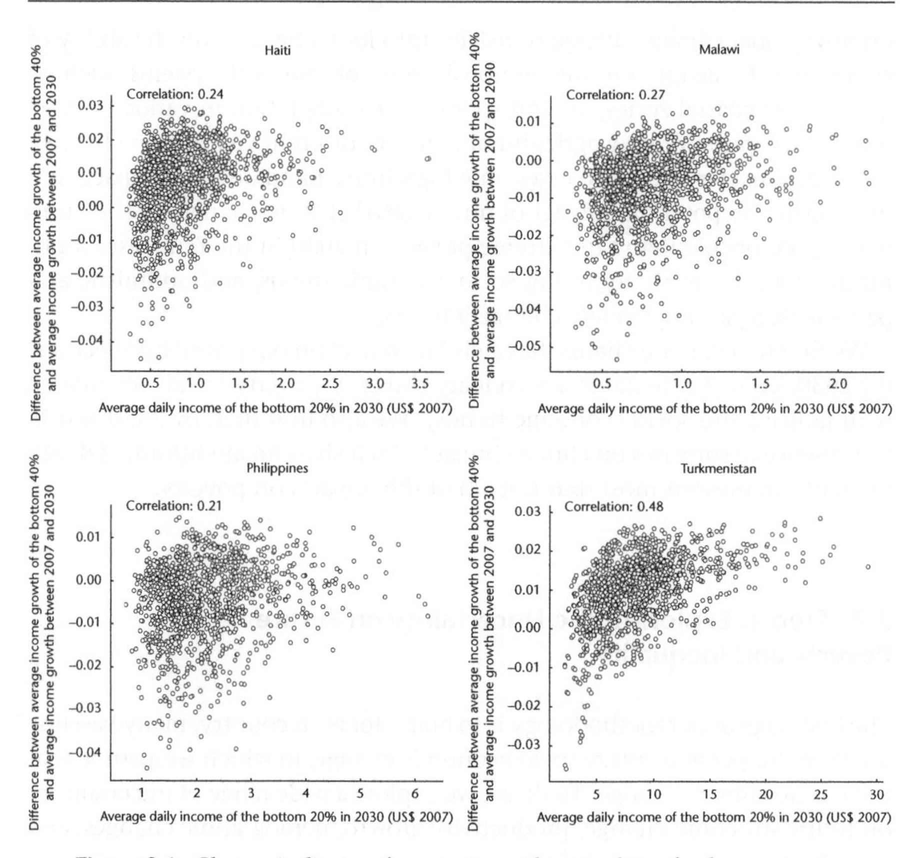

Figure 2.1. Chosen indicators for poverty and inequalities for four countries

of the bottom 40 percent and the average income growth in the country (as an indicator for inequality).

We do not attribute probabilities or likelihood to our scenarios. These scenarios thus cannot be used as forecasts or predictions of the future of poverty or as inputs in a probabilistic cost–benefit analysis. That said, they can still be an important input into decision-making as they make it possible to elicit trade-offs and highlight vulnerabilities (Kalra, Hallegatte, et al., 2014). In our case, our scenarios help us explore and quantify how poverty reduction can reduce climate change impacts.

2.3. Step 2: Selecting Two Representative Scenarios: "Prosperity" vs. "Poverty"

In each country, we select two scenarios (among the hundreds per country) that are consistent with SSP5 (with high growth and reduction of inequality) and SSP4 (with high inequality), and representative of the conditions in which poverty is reduced rapidly or more slowly (see the supplementary material in Hallegatte and Rozenberg, 2017, for details on the methods).

We then aggregate all optimistic country-level scenarios into a global optimistic scenario, labeled the *prosperity* scenario. This scenario is consistent with the World Bank twin goals of eradicating extreme poverty and promoting share prosperity: (1) the number of people living below the extreme poverty line is less than 3 percent of the global population; and (2) consumption growth for the bottom 40 percent in countries is high. We also assume that the world described in our prosperity scenario provides basic services (electricity, water, and sanitation), basic social protection, and health care and coverage to the entire population. Since each country-level scenario is chosen so that GDP growth is close to the GDP values from the SSP5 and that countries have the demographics from the SSP5, our *prosperity* scenario can be considered as a quantified pathway for poverty in the SSP5. But our *prosperity* scenario does not follow the narrative from the SSP5 concerning the energy mix and use of fossil fuels.

Similarly, we aggregate all pessimistic country-level scenarios into a global pessimistic scenario, labeled the *poverty* scenario. In this scenario, extreme poverty decreases much less, to reach 11 percent of the global population in 2030, and inequality is much larger across and within countries. In this scenario, we also assume that access to basic services, social protection, and health care improves only marginally. This scenario is consistent with the narrative from the SSP4, and our *poverty* scenario can therefore be considered a quantification of poverty in SSP4.

These two scenarios are representative of successful futures and more pessimistic ones, and can be used to assess the consequence of various shocks and stresses, accounting for the difference in vulnerability due to future socio-economic trends.

2.4. Step 3: Introducing the Impacts of Climate Change

In each country and for each of the two selected socio-economic scenarios (*prosperity* and *poverty*) we introduce climate change impacts on food price and production, natural disasters, and health. In the projections of the 1.4 million households modeled in our scenarios (representing 1.2 billion households), we adjust the income and prices to reflect the impact of climate change on their ability to consume, and derive the impact on poverty.

Given that the sector-level impacts are highly uncertain, we also define a *low-impact* and a *high-impact* scenario. These depend on the magnitude of the physical and biological impacts of climate change (which depend on the ability of ecosystems to adapt and on the responsiveness of physical systems such as glaciers and coastal zones) and on spontaneous adaptation in various sectors (such as adoption of new agricultural practices or improved hygiene habits). Note that with a 2030 horizon, impacts barely depend on emissions between 2015 and 2030, which only affect the magnitude of climate change over the longer term, beyond 2050.

There are several limits to our approach. First, we follow a bottom-up approach and sum the sectoral impacts, assuming they do not interact. Second, we consider only a subset of impacts, even within our three sectors—for instance, we do not include the loss of ecosystem services and the nutritional quality of food. Third, we cannot assess the poverty impact everywhere. Our household database represents only 83 percent of the population in the developing world. Some highly vulnerable countries (such as small islands) cannot be included in the analysis because of data limitations, in spite of the large effects that climate change could have on their poverty rates.

2.4.1. *Food Prices and Food Production*

The vulnerability of poverty reduction to food price hikes have already been demonstrated (for instance in Ivanic and Martin, 2008; Ivanic et al., 2012; Hertel et al., 2010; Devarajan et al., 2013). Impacts of climate change on agriculture affect poverty in two ways (Porter et al., 2014). First, an increase in food prices reduces households' available income, but especially consumption of the poor who spend a large share of their income on food products. The impacts in our scenarios depend on the fraction of food expenditure in total expenditure, which decreases with the income level of the household. Food price changes also affect the farmers' incomes. However, this channel is complex since lower yields mean that higher food prices do not necessarily translate into more farmers' revenues: the net effect depends on the balance between changes in prices and quantities produced.

Food prices and production come from a global agricultural model (Havlík et al., 2015). Whether higher agriculture revenues (if the increase in price dominates the decrease in yield) are transmitted to poor farmers depends on how the benefits are distributed between farm laborers and landowners (see an example on Bangladesh in Jacoby et al., 2014). To account for these effects, we assume that the increase in agriculture revenues is entirely transmitted to agriculture workers in the *prosperity* scenario (due to favorable balance of power in labor markets), but that only 50 percent is transmitted in the *poverty* scenario, the rest being captured by landowners and intermediaries in the food supply chain (who are assumed rich enough not to affect our poverty estimates).

We also rescale the (real) income of all households according to the change in food prices, accounting for the share of food in household budgets (which decreases with income). The impact of the agriculture channel on poverty depends on the number of farmers in each country, the income of these farmers, and the income of the entire population (which affects the share of food in consumption).

2.4.2. *Health*

We now include a set of additional impacts of climate change on health (stunting, malaria, and diarrhea).

2.4.2.1. STUNTING

Stunting is linked to malnutrition and therefore to food price, but acts through a different channel than the direct impact on food prices on the ability to consume. It is also driven by more than access to and affordability of food (Lloyd et al., 2011; Hales et al., 2014). Socio-economic characteristics such as parents' education and access to basic services (especially improved drinking water and sanitation) also play a key role.

Stunting has short- and long-term impacts particularly for children younger than two. For instance, households reducing nutrition after droughts permanently lowered children stature by 2.3 to 3 cm (Dercon and Porter, 2014; Alderman et al., 2006). These consequences have impacts on lifelong earning capacity and ability to escape poverty. In Zimbabwe, children affected by droughts had 14 percent lower lifetime earnings (Alderman et al., 2006) and in Ethiopia income was reduced by 3 percent for individuals who were younger than three years old during droughts.

Lloyd et al. (2011) suggest that climate change could have a large impact on stunting, and even that climate change could dominate the positive effect of development in some regions, leading to an absolute increase in stunting over time. In our model, we use the ranges given in (Hales et al. 2014, table 7.4) for the additional share of children estimated to be stunted because of climate change in 2030. We assume that stunted individuals have lifelong earning reduced by 5 percent and 15 percent in the low-impact and high-impact scenarios, respectively (regardless of their employment sector and skill level).

2.4.2.2. MALARIA

Climate change threatens to reverse the progress that has been made to date in the fight against malaria. It is difficult to identify what portion of malaria incidence can be attributed today to climate change but the World Health Report estimated climatic factors to be responsible for 6 percent of malaria cases (WHO, 2002). Further, even small temperature increases could have a great effect on transmission of malaria. At the global level, increases of 2 or 3°C could increase the number of people at risk for malaria by up to 5 percent—representing several million. Malaria could increase by 5–7 percent in populations at risk in higher altitudes in Africa, leading to an increase in the number of cases by up to 28 percent (Small et al., 2003).

In our scenarios, we use results from (Caminade et al., 2014), which give the percentage increase in malaria cases in 2030, in each country, due to climate change. Malaria is not always deadly but is a debilitating disease that often results in recurring bouts of illness (Cole and Neumayer, 2006). In this analysis, it is assumed that malaria has impacts through the cost of treatment (between $0.7 and $6 per occurrence) and lost days of work (directly or to care for someone else) (Attanasio and Székely, 1999; Konradsen et al., 1997; Ettling et al., 1994; Louis et al., 1992; Desfontaine et al., 1989; Desfontaine et al., 1990; Guiguemde et al., 1994).

2.4.2.3. DIARRHEA

As the third leading cause of death in low-income countries, diarrhea is an important risk for poor households due to easy contamination pathways resulting from unsatisfactory hygiene conditions and high exposure (WHO, 2008). Reduction in diarrhea incidence may be undermined by climate impacts that damage urban infrastructure and reduce the overall availability of water through water resource depletion.

Here we use data by Hutton et al. (2004) for the number of cases per country today and the cost of treatment (table 5). According to Kolstad and Johansson (2010) the prevalence of diarrhea could increase by 10 percent by 2030 because of climate change (in all regions), and we use this assessment in our scenarios.

Further, we assume that fast progress in access to water and sanitation in the *prosperity* scenario halve the number of cases, consistently with the assessment in India by (Andres et al., 2014). Of course, this assumes that new water and sanitation infrastructure are adapted to changing future climate conditions and can continue to perform well in 2030 and beyond. This would be required to account for the uncertainty in climate projections in the design phase, and to invest in the additional cost of more resilient infrastructure, possibly factoring in safety margins and retrofit options (Kalra, Gill, et al. 2014).

2.4.3. *Temperature and Labor Productivity*

Recent studies suggest that there is a significant impact of temperature stress on labor productivity, which may be exacerbated by global climate change (Dell et al., 2014). In particular, there are direct physiological effects of thermal stress on the human body, which may affect productivity and labor supply, especially in developing countries (Heal and Park, 2015). Using variations in weather, several studies identified a relationship between extreme temperature—for instance, hotter-than-average years or extremely hot days—and economic outcomes such as labor productivity (Burke et al., 2015; Sudarshan et al., 2014; Dell et al., 2014). Park et al. (2015) suggests the existence of an optimal temperature for economic activity, based on an analysis of non-agricultural payroll in US counties between 1986 and 2012.

Today, temperatures are about 1°C higher than they would be in the absence of climate change. In 2030, the difference will be around 1.2 or 1.4°C. In our analysis, we assume that people working outside or without air conditioning will lose between 1 and 3 percent in labor productivity due to this change of climate, compared with a baseline with no climate change.

2.4.4. *Natural Disasters*

For natural disasters, we work based on orders of magnitude using the EM-DAT database, and focus on direct economic losses, disregarding human losses and indirect and second-order losses (Hallegatte, 2012, 2014). We start from current economic losses due to natural disasters, which have evolved between $50 and $200 billion in recent years, i.e. between 0.05 and 0.2 percent of world GDP.

We assume that the disasters in the no climate change scenarios are already included in the baseline socio-economic scenario, and we add to our simulations the additional disaster losses due to climate change. We make crude assumptions on how climate change will affect disaster losses, reflecting the large uncertainty of the effect of climate change on extreme events and the fact that losses will be highly dependent on how protections and other adaptation measures change over time.

2.5. Results: Impacts of Climate Change on Poverty

The impact of climate change on agricultural production is the chief culprit in all four scenarios (*prosperity* and *poverty*, combined with high and low impacts) (Figure 2.2). Next come health impacts (diarrhea, malaria, and stunting) and the labor productivity effects of high temperature with a second-order but significant role. Disasters have a limited impact in our simulations, but we have to remain careful because only the direct impact on income losses was taken into account.

Agriculture is the channel through which climate change has the biggest impact on poverty because the most severe food price increase and reduction

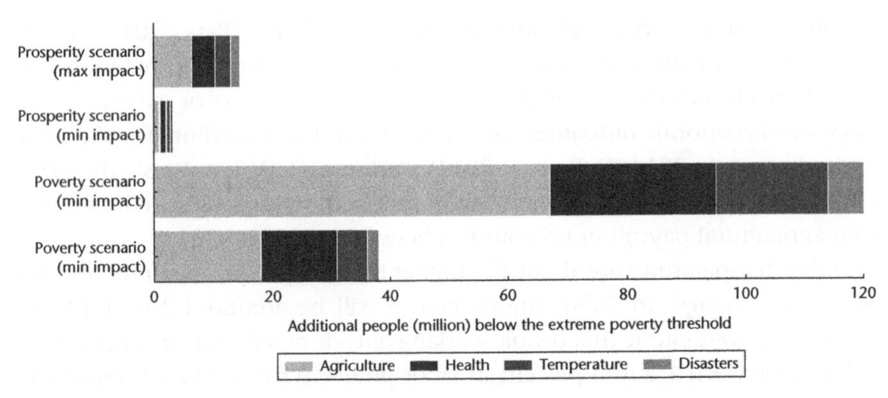

Figure 2.2. Summary of climate change impacts on the number of people living below the extreme poverty threshold, by source. Agriculture is the main sectoral factor explaining higher poverty due to climate change

in food production happen in sub-Saharan Africa and India, where most poor people live in 2030.

How do these sectoral results add up in terms of climate change's effect on future poverty trends? We find that a large effect on poverty is possible, even though our analysis is partial and does not include many other possible impacts (for example through tourism and energy prices) and looks only at the short term (during which there will be small changes in climate conditions compared with what unabated climate change could bring over the long term). Indeed, our overall results show that between 3 million and 122 million additional people would be in poverty because of climate change in our main *prosperity* and *poverty* scenarios (Table 2.1):

- In the *poverty* scenario, the total number of people living below the extreme poverty line in 2030 is 1.02 billion in the high-impact scenario; this represents an increase of 122 million people compared to a scenario with no climate change. For the low-impact scenario, the additional number of poor people is 35 million.
- In the *prosperity* scenario, the increase in poverty due to a high-impact climate change scenario is "only" 16 million people, suggesting that development and access to basic services (like water and sanitation) is effective in reducing poor people's vulnerability to climate change. For the low-impact scenario, the additional number of poor people is 3 million.

Note that the large range of estimates in our results—3 to 122 million—may incorrectly suggest that we cannot say anything about the future impact of climate change on poverty. The reason for this rather wide range is not just scientific uncertainty on climate change and its impacts. Instead, it is predominantly policy choices—particularly those concerning development

Table 2.1. Impact of climate change on extreme poverty

		Climate change scenario				
		No climate change	Low-impact scenario		High-impact scenario	
		Number of people in extreme poverty	Additional number of people in extreme poverty due to climate change			
Socio-economic scenario	Prosperity scenario	142 million	+3 million		+16 million	
			Minimum +3 million	Maximum +6 million	Minimum +16 million	Maximum +25 million
	Poverty scenario	899 million	+35 million		+122 million	
			Minimum −25 million	Maximum +97 million	Minimum +33 million	Maximum +165 million

Note: The main results use the two representative scenarios for prosperity and poverty. The ranges are based on the sixty alternative scenarios for each category (10–90 percentiles). These simulations are performed using 2005 PPP exchange rate and the $1.25 extreme poverty line, but results are not expected to change significantly under the $1.90 poverty line and using 2011 PPP.

patterns and poverty reduction policies between now and 2030. While emissions reduction policies cannot do much regarding the climate change that will happen between now and 2030 (since that is mostly the result of past emissions), development choices can affect what the impact of that climate change will be.

In the *prosperity* scenario, the lower impact of climate change on poverty comes from a reduced vulnerability of the developing world to climate change compared to the *poverty* scenario. This reduced vulnerability, in turn, stems from several channels:

- People are richer and fewer households live with a daily income close to the poverty line. Wealthier people are less exposed to health shocks (such as stunting and diarrhea) and are less likely to be pushed into poverty when hit by a shock.
- The global population is smaller in the *prosperity* scenario in 2030, by 2 percent globally, 4 percent in the developing world, and 10 to 20 percent in most African countries. This difference in population makes it easier for global food production to meet demand, thereby mitigating the impact of climate change on global food prices. The *prosperity* scenario also assumes more technology transfers to developing countries, which further mitigates agricultural losses.
- There is more structural change (involving shifts from unskilled agricultural jobs to skilled manufacturing and service jobs), so fewer workers are vulnerable to the negative impacts of climate change on yields. In the *prosperity* scenario, a more balanced economy and better governance mean that farmers capture a larger share of the income benefits from higher food prices.

a. Prosperity scenario

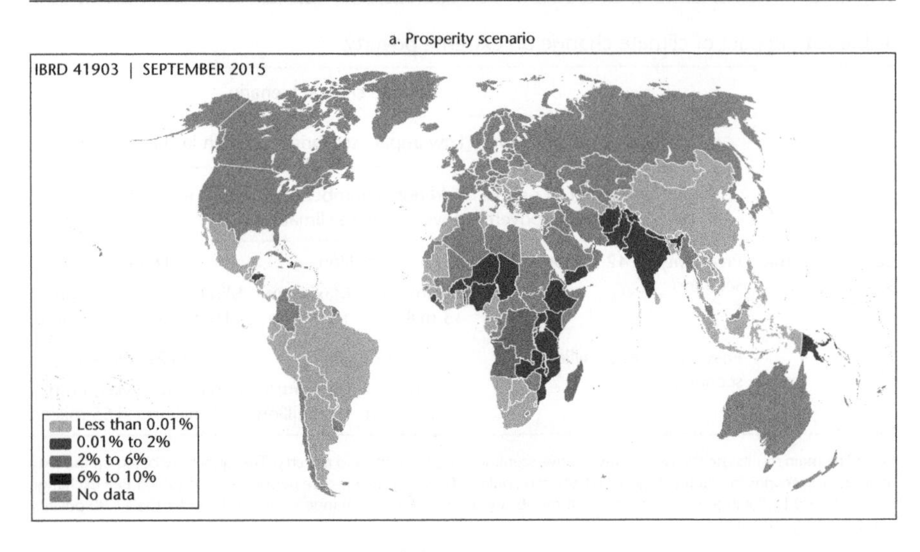

b. Poverty scenario

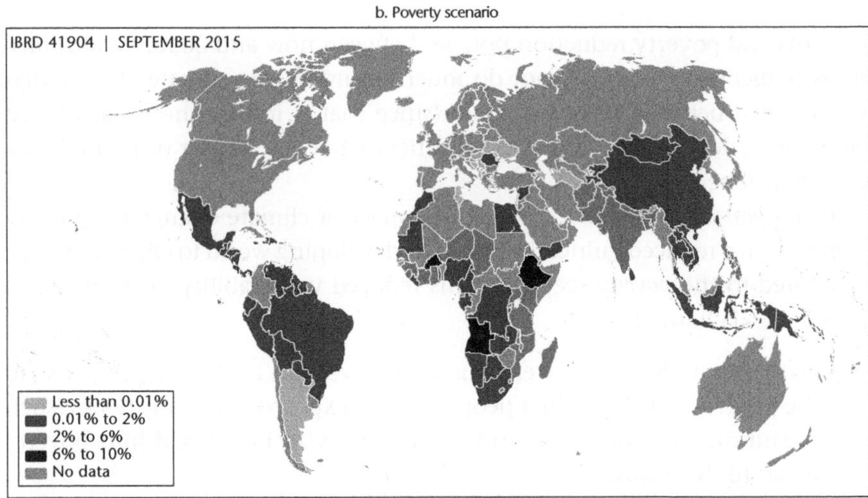

Figure 2.3. Increase in poverty rate due to climate change in the worst case climate change scenario considered

Up to 2030, climate change remains a secondary driver of global poverty compared to development: the difference across reference scenarios due to socio-economic trends and policies (that is, the difference between the *poverty* and *prosperity* scenarios in the absence of climate change) is almost 800 million people. This does not mean that climate change impacts are secondary at the local scale: in some particularly vulnerable places (like small islands or in unlucky locations affected by large disasters), the local impact could be massive.

Our global results in the representative *prosperity* and *poverty* scenarios also hide higher impacts at a finer scale. At the country and regional level, the hotspots for increased poverty because of climate change are sub-Saharan Africa and—to a lesser extent—India and the rest of South Asia, especially in the *poverty* scenario (Figure 2.3). Those countries, in Africa in particular, bear a higher burden because they have the highest initial number of poor people and the steepest projected food price increases.

In almost all countries, the additional number of poor people due to climate change is higher in the *poverty* scenario than in the *prosperity* scenario. Two exceptions are Liberia and the Democratic Republic of the Congo, for which the number of poor people pushed into poverty because of climate change is higher in the *prosperity* scenario than in the *poverty* scenario. This is because, in the *poverty* scenario, 70 percent of the population still lives below the extreme poverty threshold in 2030 even without climate change. There are fewer people at risk of falling into poverty because most of the population is already poor—a reminder that the depth of poverty (not just the poverty headcount) also matters.

Finally, we find that impacts of the poor are likely to be larger than the average impact, even within countries. Figure 2.4 plots the income losses at the aggregate level (x-axis) and for the "bottom 40 percent," i.e. the 40 percent of households with the lowest income in the country (these metrics build on the "shared prosperity" metrics used by the World Bank; see World Bank and International Monetary Fund 2014). Figure 2.4 mixes results for all countries, and for the four scenarios. The impacts on the bottom 40 percent are 60 percent are larger than those of the average population.

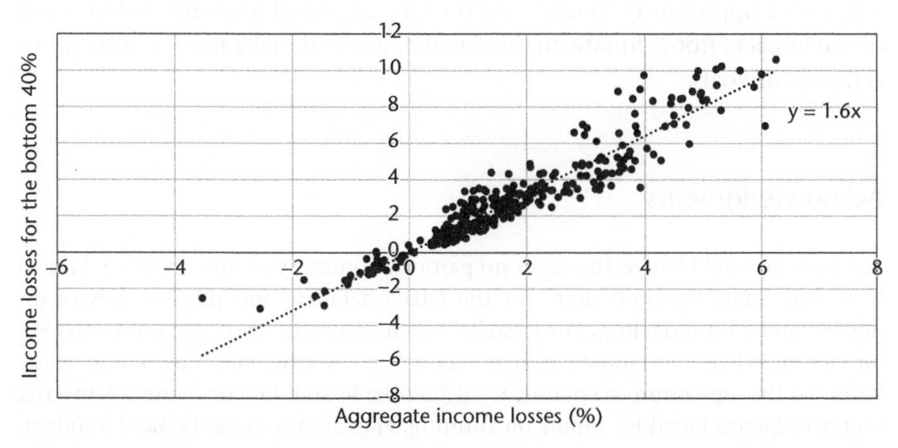

Figure 2.4. Income losses for the poorest 40 percent in each country and in our four scenarios (y-axis) against average income losses

2.6. Limits and Conclusion

The numbers presented here should not be taken as forecasts or predictions. We have built two development scenarios among hundreds of possibilities and calculated highly uncertain impacts of climate change in those two scenarios. Also, the impact of climate change will depend on many policies and development trends (such as the access of developing countries to world food markets or innovation in crops and agricultural techniques) that are not represented in our modeling framework.

However, a few robust insights can be drawn from this exercise. The quantitative impacts of climate change on poverty are uncertain, but they are likely to be significant, even over the relatively short term. Our analysis only covers a small fraction of all climate change impacts—for instance it does not account for the impact on ecosystem services—but still finds that 120 million people may be trapped in poverty because of climate change impacts. By 2030, however, climate change remains a secondary driver of poverty, and is less important than demographics, socio-economic factors, and policy changes. We cannot conclude whether this remains valid over the longer term.

The quantitative impacts of climate change on poverty are much smaller in a world where socio-economic trends and policies ensure that development is rapid, inclusive, and climate-informed than in a world where extreme poverty would persist even without climate change. Development policies therefore appear to be good adaptation policies. Climate change is, however, creating a renewed urgency: if poverty is not reduced rapidly, then the impact of climate change will make it even more difficult to eradicate poverty later. We have a window of opportunity to eradicate poverty and build resilience before most of the impacts from climate change materialize and make it more difficult to achieve our goals.

Acknowledgments

The authors would like to thank, in no particular order, Syud Amer Ahmed, Marcio Cruz, and Israel Osorio-Rodarte for the GIDD database and precious advice on micro-simulation modeling; Jan Kwakkel for the scenario discovery tools; Kris Ebi for her knowledge of climate change impacts on health; Petr Havlík and Hugo Valin for the agriculture scenarios; Cyril Caminade and Tamaro Kane for malaria numbers; Simon Lloyd for inputs on stunting; Francisco Ferreira, Eeshani Kandpal, Marianne Fay, and Adrien Vogt-Schilb for comments on previous versions of this chapter.

References

Ahmed, Syud A., Noah S. Diffenbaugh, and Thomas W. Hertel. 2009. "Climate Volatility Deepens Poverty Vulnerability in Developing Countries." *Environmental Research Letters* 4 (3): 034004.

Alderman, Harold, John Hoddinott, and Bill Kinsey. 2006. "Long Term Consequences of Early Childhood Malnutrition." *Oxford Economic Papers* 58 (3): 450–74.

Andres, Luis, Bertha Briceño, Claire Chase, and Juan A. Echenique. 2014. "Sanitation and Externalities: Evidence from Early Childhood Health in Rural India." World Bank Policy Research Working Paper, no. 6737. <http://papers.ssrn.com/sol3/papers.cfm?abstract_id=2375456>.

Attanasio, Orazio, and Miguel Székely. 1999. "An Asset-Based Approach to the Analysis of Poverty in Latin America." SSRN Scholarly Paper ID 1814653. Rochester, NY: Social Science Research Network. <http://papers.ssrn.com/abstract=1814653>.

Bourguignon, François, Francisco HG Ferreira, and Nora Lustig. 2005. *The Microeconomics of Income Distribution Dynamics in East Asia and Latin America*. Washington, DC: World Bank Publications.

Burke, M., S. M. Hsiang, and E. Miguel. 2015. "Global Non-linear Effect of Temperature on Economic Production." *Nature* 527 (7577): 235–9.

Bussolo, Maurizio, Rafael De Hoyos, Denis Medvedev, et al. 2008. "Economic Growth and Income Distribution: Linking Macroeconomic Models with Household Survey Data at the Global Level." International Association for Research in Income and Wealth (IARIW) 30th General Conference, Portoroz, Slovenia, 24–30 August. <http://siteresources.worldbank.org/INTPROSPECTS/Resources/334934-1225141925900/Buss%26DeH%26Med2008LinkingHHSurveyDataIARIW.pdf>.

Caminade, Cyril, Sari Kovats, Joacim Rocklov, Adrian M. Tompkins, Andrew P. Morse, Felipe J. Colón-González, Hans Stenlund, Pim Martens, and Simon J. Lloyd. 2014. "Impact of Climate Change on Global Malaria Distribution." *Proceedings of the National Academy of Sciences* 111 (9): 3286–91.

Cole, Matthew A., and Eric Neumayer. 2006. "The Impact of Poor Health on Total Factor Productivity." *The Journal of Development Studies* 42 (6): 918–38.

Dell, M., B. F. Jones, and B. A. Olken. 2009. "Temperature and Income: Reconciling New Cross-Sectional and Panel Estimates." *American Economic Review* 99: 198–204.

Dercon, Stefan, and Catherine Porter. 2014. "Live Aid Revisited: Long-Term Impacts of the 1984 Ethiopian Famine on Children." *Journal of the European Economic Association* 12 (4): 927–48.

Desfontaine, M., H. Gelas, H. Cabon, A. Goghomou, D. Kouka Bemba, and P. Carnevale. 1990. "Evaluation of Practice and Costs of Vector Control on a Family Level in Central Africa. II. Douala City (Cameroon), July 1988." *Annales De La Société Belge De Médecine Tropicale* 70 (2): 137–44.

Desfontaine, M., H. Gelas, A. Goghomu, D. Kouka-Bemba, and P. Carnevale. 1989. "Evaluation of Practices and Costs of Antivectorial Control at the Family Level in Central Africa, I. Yaoundé City (March 1988)." *Bulletin De La Société De Pathologie Exotique Et De Ses Filiales* 82 (4): 558–65.

Devarajan, Shantayanan, S. Delfin Go, Maryla Maliszewska, Israel Osorio-Rodarte, and Hans Timmer. 2013. "Stress-Testing Africa's Recent Growth and Poverty Performance." Policy Research Working Paper 6517, Development Prospect Group and Africa Region, Office of the Chief Economist, World Bank.

Ettling, M., D. A. McFarland, L. J. Schultz, and L. Chitsulo. 1994. "Economic Impact of Malaria in Malawian Households." *Tropical Medicine and Parasitology: Official Organ of Deutsche Tropenmedizinische Gesellschaft and of Deutsche Gesellschaft Für Technische Zusammenarbeit (GTZ)* 45 (1): 74–9.

Guiguemde, T. R., F. Dao, V. Curtis, A. Traore, B. Sondo, J. Testa, and J. B. Ouedraogo. 1994. "Household Expenditure on Malaria Prevention and Treatment for Families in the Town of Bobo-Dioulasso, Burkina Faso." *Transactions of The Royal Society of Tropical Medicine and Hygiene* 88 (3): 285–7.

Gupta, Monica Das. 2014. "Population, Poverty, and Climate Change." *The World Bank Research Observer* 29 (1): 83–108.

Hales, Simon, Sari Kovats, Simon Lloyd, and Diarmid Campbell-Lendrum. 2014. "Quantitative Risk Assessment of the Effects of Climate Change on Selected Causes of Death, 2030s and 2050s." World Health Organization. <http://www.who.int/globalchange/publications/quantitative-risk-assessment/en/>.

Hallegatte, S. 2012. "Modeling the Roles of Heterogeneity, Substitution, and Inventories in the Assessment of Natural Disaster Economic Costs." World Bank Policy Research Working Paper, no. 6047. <http://papers.ssrn.com/sol3/papers.cfm?abstract_id=2045387>.

Hallegatte, Stephane. 2014. "Economic Resilience: Definition and Measurement." Policy Research Working Paper 6852, World Bank.

Hallegatte, Stephane, Valentin Przyluski, and Adrien Vogt-Schilb. 2011. "Building World Narratives for Climate Change Impact, Adaptation and Vulnerability Analyses." *Nature Climate Change* 1 (3): 151–5.

Hallegatte, Stephane, and Julie Rozenberg. 2017. "Climate Change Through a Poverty Lens." *Nature Climate Change* 7 (4): 250–6.

Havlík, Petr, Hugo Valin, Mykola Gusti, Nicklas Forsell, Mario Herrero, David Leclère, Nikolay Khabarov, Aline Mosnier, Michael Obersteiner, and Erwin Schmid. 2015. "Climate Change Impacts and Mitigation in the Developing World: Integrated Assessment of Agriculture and Forestry Sectors." Forthcoming as a World Bank Policy Research Working Paper.

Heal, G., and J. Park. 2013. "Feeling the Heat: Temperature, Physiology and the Wealth of Nations." Working Paper No. 19725, National Bureau of Economic Research.

Hertel, Thomas W., Marshall B. Burke, and David B. Lobell. 2010. "The Poverty Implications of Climate-Induced Crop Yield Changes by 2030." *Global Environmental Change*, 20th Anniversary Special Issue, 20 (4): 577–85.

Hertel, Thomas W., and Stephanie D. Rosch. 2010. "Climate Change, Agriculture, and Poverty." *Applied Economic Perspectives and Policy* 32 (3): 355–85.

Hutton, Guy, Laurence Haller, et al. 2004. "Evaluation of the Costs and Benefits of Water and Sanitation Improvements at the Global Level." Water, Sanitation, and Health, Protection of the Human Environment, World Health Organization. <http://wwwlive.who.int/entity/water_sanitation_health/wsh0404.pdf>.

IPCC. 2013. "Summary for Policymakers." In *Climate Change 2013: The Physical Science Basis*, edited by T. F. Stocker, D. Qin, G.-K. Plattner, M. Tignor, S. K. Allen, J. Boschung, A. Nauels, Y. Xia, V. Bex, and P. M. Midgley, 1–30. Contribution of Working Group I to the Fifth Assessment Report of the Intergovernmental Panel on Climate Change. Cambridge and New York: Cambridge University Press.

Ivanic, Maros, and Will Martin. 2008. "Implications of Higher Global Food Prices for Poverty in Low-Income countries." *Agricultural Economics* 39 (November): 405–16.

Ivanic, Maros, Will Martin, and Hassan Zaman. 2012. "Estimating the Short-Run Poverty Impacts of the 2010–11 Surge in Food Prices." *World Development* 40 (11): 2302–17.

Jacoby, Hanan G., Mariano Rabassa, and Emmanuel Skoufias. 2014. "Distributional Implications of Climate Change in Rural India: A General Equilibrium Approach." *American Journal of Agricultural Economics*, 97 (4): 1135–56.

Kalra, Nidhi, Stuart Gill, Stéphane Hallegatte, Casey Brown, Adrian Fozzard, Robert Lempert, and Ankur Shah. 2014. "Agreeing on Robust Decisions: New Processes for Decision Making under Deep Uncertainty." WPS6906. The World Bank, Washington, DC. <http://documents.worldbank.org/curated/en/2014/06/19616379/agreeing-robust-decisions-new-processes-decision-making-under-deep-uncertainty>.

Kolstad, Erik W., and Kjell Arne Johansson. 2010. "Uncertainties Associated with Quantifying Climate Change Impacts on Human Health: A Case Study for Diarrhea." *Environmental Health Perspectives* 119 (3): 299–305.

Konradsen, F., W. van der Hoek, P. H. Amerasinghe, F. P. Amerasinghe, and K. T. Fonseka. 1997. "Household Responses to Malaria and Their Costs: A Study from Rural Sri Lanka." *Transactions of The Royal Society of Tropical Medicine and Hygiene* 91 (2): 127–30.

Lloyd, Simon J., R. Sari Kovats, and Zaid Chalabi. 2011. "Climate Change, Crop Yields, and Undernutrition: Development of a Model to Quantify the Impact of Climate Scenarios on Child Undernutrition." *Environmental Health Perspectives* 119 (12): 1817–23.

Louis, J. P., A. Trebucq, H. Gelas, E. Fondjo, L. Manga, J. C. Toto, and P. Carnevale. 1992. "Malaria in Yaounde (Cameroon): Cost and Antivectorial Control at the Family Level." *Bulletin De La Société De Pathologie Exotique* 85 (1): 26–30.

Olivieri, Sergio, Sergiy Radyakin, Stanislav Kolenikov, Michael Lokshin, Ambar Narayan, and Carolina Sanchez-Paramo. 2014. *Simulating Distributional Impacts of Macro-Dynamics: Theory and Practical Applications*. Washington, DC: World Bank Publications.

O'Neill, Brian C., Elmar Kriegler, Keywan Riahi, Kristie L. Ebi, Stéphane Hallegatte, Timothy R. Carter, Ritu Mathur, and Detlef P. van Vuuren. 2013. "A New Scenario Framework for Climate Change Research: The Concept of Shared Socioeconomic Pathways." *Climatic Change* 122 (3): 387–400.

Park, J., Stéphane Hallegatte, Mook Bangalore, and Evan Sandhoefner. 2015. "The Deck Is Stacked (and Hot)? Climate Change, Labor Productivity, and Developing Countries," World Bank Policy Research Working Paper No. 7479.

Porter, J. R., L. Xie, A. J. Challinor, K. Cochrane, S. M. Howden, M. M. Iqbal, D. B. Lobell, and M. I. Travasso. 2014. "Food Security and Food Production Systems." In C. B. Field, V. R. Barros, D. J. Dokken, K. J. Mach, M. D. Mastrandrea, T. E. Bilir, M. Chatterjee, et al. (eds.), *Climate Change 2014: Impacts, Adaptation, and Vulnerability*.

Part A: Global and Sectoral Aspects. Contribution of Working Group II to the Fifth Assessment Report of the Intergovernmental Panel of Climate Change, 485–533. Cambridge and New York: Cambridge University Press.

Rozenberg, Julie, and Stéphane Hallegatte. 2015. "The Impacts of Climate Change on Poverty in 2030, and the Potential from Rapid, Inclusive and Climate-Smart Development." Forthcoming as a World Bank Policy Research Working Paper.

Skoufias, Emmanuel, Mariano Rabassa, and Sergio Olivieri. 2011. "The Poverty Impacts of Climate Change: A Review of the Evidence." SSRN Scholarly Paper ID 1803002. Social Science Research Network, Rochester, NY. <http://papers.ssrn.com/abstract=1803002>.

Small, J., S. J. Goetz, and S. I. Hay. 2003. "Climatic Suitability for Malaria Transmission in Africa, 1911–1995." *Proceedings of the National Academy of Sciences* 100: 15, 341–45.

Sudarshan, A., E. Somanathan, R. Somanathan, and M. Tewari. 2015. "The Impact of Temperature on Productivity and Labor Supply: Evidence from Indian Manufacturing." Working Paper No 244, Centre for Development Economics, Delhi School of Economics.

WHO (World Health Organization). 2002. *The World Health Report.* Geneva: World Health Organization.

WHO. 2008. *Global Burden of Disease: 2004 Update.* Geneva: World Health Organization.

World Bank, and International Monetary Fund. 2014. *Global Monitoring Report 2014/2015: Ending Poverty and Sharing Prosperity.* World Bank Publications.

3

Governing the Commons to Promote Global Justice: Climate Change Mitigation and Rent Taxation

Michael Jakob, Ottmar Edenhofer, Ulrike Kornek, Dominic Lenzi, and Jan Minx

3.1. The Challenge of Managing Global Commons

The atmosphere is a global common pool resource. It is *de jure* a "res nullius" with open access as a disposal space for greenhouse gases (GHGs). Guaranteeing the sustainable and equitable use of the atmosphere requires the establishment of a global common property regime. Even though the voluntary national mitigation efforts agreed upon in the Paris Agreement are not legally binding, they have established a target to limit global temperature increase "well below 2°C" (UNFCCC, 2015). The Paris Agreement might also enable nation states to cooperate and coordinate their national efforts toward a global common property regime for the atmosphere.

It is well understood that justice is central to climate change mitigation (Gardiner et al., 2010). People in developing countries will be most affected by climate change, whereas the largest share of GHGs in the atmosphere has been emitted in industrialized countries (Jakob and Steckel, 2014). As the most severe climate impacts can be expected to occur in the far future, climate policy also entails important intergenerational equity dimensions (Markandya, 2011).

Restricting the use of the atmosphere would create a novel scarcity rent, namely the "climate rent." Climate policy would reduce the rents accruing to the owners of coal, oil, and gas—as demand for these fossil fuels declines, so would their value. Climate change mitigation policies hence bear important implications for global justice, not only in terms of avoiding impacts, but also concerning how the climate rent is distributed. Scarcity rents accrue for the atmosphere (for which scarcity is created as a result of an agreement to restrict its use), but also for land and natural resources, which face natural limitations. Although land and natural resources are usually governed on national or subnational levels, just distribution of these rents will become an important issue.

This chapter argues that climate, resource, and land rents could be used to promote sustainable socio-economic development, for instance by investing in education, health, or basic infrastructure. The Agenda 2030 to achieve the Sustainable Development Goals (SDGs) requires substantial additional funds, which will to a large extent need to be mobilized domestically (Franks et al., n.d.). We explore the potential of rent taxation for financing investments in SDGs. The trade-off between economic and social development and environmental integrity should be studied within this broader perspective. This would allow for a better understanding of how to design policies that provide an appropriate balance between short-term socio-economic and long-term environmental targets (Jakob and Edenhofer, 2014).

This chapter proceeds as follows: Section 3.2 discusses how the climate rent is created, which ethical arguments have been brought forth regarding its distribution, and how it could contribute to financing sustainable socio-economic development. Section 3.3 analyzes these issues for land rents as well as natural resource rents. Section 3.4 discusses the political economy rent taxation and implementation issues. Section 3.5 concludes.

3.2. The Climate Rent

Limiting the use of the atmosphere will create novel property rights. Carbon pricing can be welfare enhancing for two reasons (Edenhofer et al., 2015). First, it internalizes a global externality and can provide additional benefits, such as reducing local air pollution. Second, the associated public revenues can be used to lower distortionary taxes or to address underinvestment in public goods. To meet the SDG financing needs, public spending will need to be increased in most developing countries, which will require additional tax revenues. As due to institutional constraints (such as lack of administrative capacity and a large informal sector) possibilities to raise existing taxes remain severely restricted, taxing GHG emissions could be a promising source of additional public revenues. This section discusses what could be achieved by using fiscal instruments (e.g. taxes or tradable permit schemes) to put a price on using the atmosphere and using the associated public revenues to advance human well-being.

3.2.1. *Implementation of Carbon Pricing on the International Level*

Every ton of GHGs entering the atmosphere changes the global climate, irrespective of where it is emitted. The atmosphere therefore constitutes a global commons: on the one hand, the ability of the atmosphere to take up emissions is limited and overuse will lead to additional climate impacts. On the other hand, under current international law no country can be forced to stop using the atmosphere as a sink for its GHG emissions. International cooperation is required, in which countries are willing to voluntarily constrain their national emissions in mutual agreement with every other country. In 2015, nearly all countries agreed to cooperate under the Paris Agreement to enforce international emission reductions. The new global climate governance architecture relies on three main pillars. First, countries established the global goal to keep temperature increase well below 2°C, with a view to limit warming further to 1.5°C. The 2°C target limits the disposal space of the atmosphere for CO_2 emissions to roughly 800 $GtCO_2$ by the end of the century (Edenhofer et al., 2016). At the current rate (of about 40 $GtCO_2$ per year), this "carbon budget" will be used up in approximately two decades. Second, the agreement obliges all parties to submit an emissions reduction target in their Nationally Determined Contribution (NDC). These voluntary pledges are not based on a shared distribution of the global carbon budget allowed for reaching the 2°C target. Instead, each country establishes its own climate policy ambition and the planned efforts of all NDCs are aggregated and compared to objectives. Countries are then asked to increase their ambition level in a pledge-and-review system if the global emissions reduction level falls short of what is required to stay within the temperature targets. The third pillar constitutes a set of multilateral climate policy instruments used for distributing the burden among the members of the international community and for facilitating cooperation between countries.

International cooperation within the Paris Agreement is impeded by the global commons nature of the atmosphere and associated free-riding incentives (Lessmann et al., 2015). Ambitious emission reductions of an individual country are only effective in achieving the global goal of limiting warming to well below 2°C if other states reciprocate the effort and globally aggregate emissions remain within the carbon budget. However, countries can reduce their own mitigation costs by lowering the ambition of their NDC, free-riding on the efforts of other countries. Concerns about national competitiveness and differences in the costs of emission reductions are further obstacles to ambitious global climate protection. Appropriate institutions and policy instruments are needed to counteract free-riding and increase ambition over time (Barrett, 2005). Currently, the aggregate effort of intended NDCs is inconsistent with the ambitious targets of the Paris Agreement (Edenhofer et al., 2016). Complementary development of the pledge-and-review system is required.

There are several reasons why negotiations over internationally coordinated carbon prices are a promising way to enhance climate policy ambition

(MacKay et al., 2015; Edenhofer et al., 2016). First, as will be argued in more detail later on, carbon prices are an effective policy instrument to achieve emissions reductions of NDCs at the least cost. Second, carbon prices are easy to compare. A price on carbon is an approximate indicator of the climate policy ambition and abatement costs of an individual country. While the same absolute emission reduction target can imply very different efforts—dependent on the reference level and stage of development of a country—a higher carbon price always leads to additional emission reductions. Negotiations over carbon prices therefore allow for conditional commitments to establish reciprocity between countries. An increase in the individual effort of a country through a higher national carbon price will only be realized if other countries do the same. In turn, if a country lowers their carbon price, other countries do the same, such that environmental quality on the whole is lowered as a punishment for the defector.

Negotiations about carbon prices need to respect the "common but differentiated responsibilities" principle of the UNFCCC. Coordination on a common carbon price could start with a smaller set of countries that are at a stage of development allowing for the implementation of such a policy. The G20 could serve as an effective forum to start these negotiations. However, reaching the temperature goals cooperatively implies that other countries implement carbon prices at a comparable level in the future. In this case rich and poor countries must agree to a burden-sharing mechanism. Using the $100 billion of climate funding mobilized through the Paris Agreement, transfer payments could be made to poorer countries conditional on their minimum price for emissions (Cramton et al., 2015). The resources for example in the Green Climate Fund would then link to the climate policy ambition of an individual country (Kornek and Edenhofer 2016). A country with a comparatively high carbon price would be compensated for its higher abatement costs, which creates an incentive for this country to pursue more ambitious climate policies in its NDC. This system has the potential to overcome free-riding incentives as it facilitates reciprocity between heterogeneous countries.

3.2.2. *Carbon Pricing Revenues*

Realizing the emission reductions inscribed in the NDCs will be pivotal to the success of the Paris Agreement. While the agreement itself does not specify how each country should reduce emissions, many NDCs include carbon pricing as a climate policy instrument. A carbon price makes CO_2 emissions costly and implements the polluter pays principle. In particular, high-emission forms of production (e.g. energy from coal fired power plants) become more expensive and, if the price is sufficiently high, unprofitable over the long term. At the same time, renewable energies such as wind and solar power become more competitive. As carbon pricing reduces emissions irrespective of their origin, it is a promising policy instrument to achieve a country's NDC in a cost-efficient manner. A country can put a price on carbon by using an emissions trading scheme (for example in the European Union), a tax on emissions (as done in Sweden), or combining both instruments in a hybrid system (as in the Californian emissions trading scheme with a minimum price). Over the years, more countries have begun to use carbon pricing and more schemes are scheduled for the coming years. Today, existing regional, national, or subnational carbon pricing schemes cover about 12 percent of global emissions (Kossoy et al., 2015). If China implements its envisaged economy-wide emission trading system, this share can be expected to rise to more than 20 percent.

One of the most crucial questions regarding carbon pricing concerns the appropriate price level. Two approaches are commonly used. First, the idea of internalizing an environmental externality requires estimating the "social costs of carbon" (SCC), i.e. the economic damages arising from emitting one unit of GHG. Estimating the SCC is notoriously contentious, not only due to substantial uncertainties regarding physical climate impacts, their economic valuation, and possibilities for adaptation, but also due to the inherent normative characteristics of deciding how damages that occur to future generations should be appropriately accounted for. A second approach is to determine carbon prices that would be required to achieve a certain stabilization target, such as 2°C, without passing a verdict on whether this target is socially optimal. The resulting prices crucially depend on numerous parameters, such as the stabilization target, specific model characteristics, the availability of key technologies, and the participation of key emitters in a global mitigation effort. For instance, a recent high-level commission has recommended carbon prices of between $40 and $80 for the year 2020 to ensure a reasonable chance of achieving the 2°C target (Carbon Pricing Leadership Coalition 2017).

Some object that the very aim of carbon pricing is to reduce emissions, i.e. to erode its own tax base, therefore making revenues from carbon pricing difficult to support as an important source for the public budget. This argument fails to take into account the inverse relationship between the carbon price and emissions, which can result in either increasing or decreasing revenues (which are determined by multiplying emissions with the carbon price) depending on how sensitively emissions react to a carbon price. This sensitivity is an empirical issue that is mainly determined by the inertia of the energy system. A comparison of seven integrated energy-climate-economy models (Blanford et al., 2014, p. 27; Kriegler et al., 2014) reveals that carbon pricing consistent with a 2°C climate target would in fact result in rising revenues (despite declining emissions) until the year 2040, as shown in Figure 3.1.

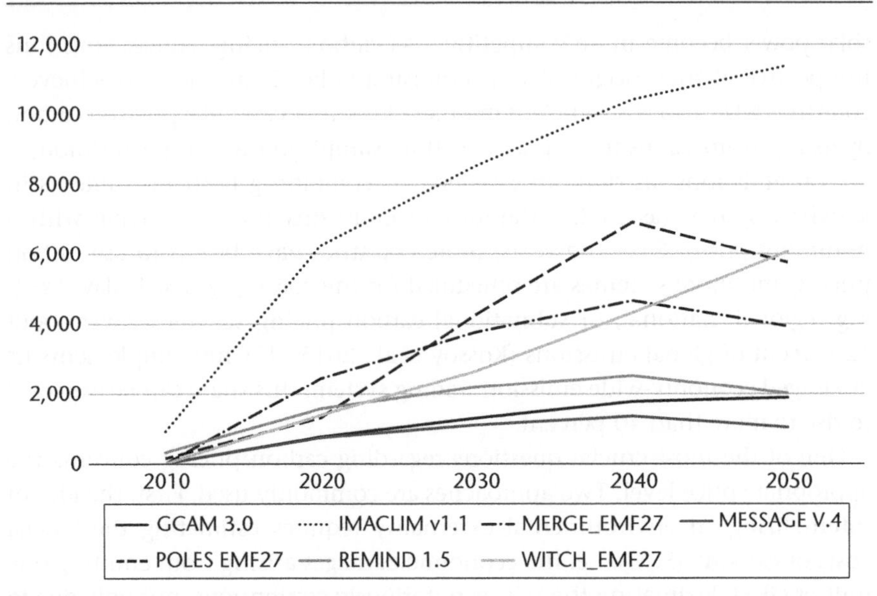

Figure 3.1. Revenues from carbon pricing under a 450 ppm scenario with full technological availability for seven integrated assessment models

Source: Data from IIASA (2014), based on scenarios from the EMF27 model comparison (Blanford et al. 2014; Kriegler et al. 2014).

Hence, in a setting with ambitious climate change mitigation polices, carbon pricing could be counted on as an important revenue source over the period of at least several decades.

3.2.3. *Just Distribution and Revenue Recycling on the International Level*

The creation of a new stream of revenues from carbon pricing and resource rent recycling raises questions concerning their just or fair distribution. Theories of justice help inform what a just distribution of these revenues might look like. But one must ask not only how revenues ought to be distributed within states, but whether some states ought to redistribute a proportion of their revenues to poorer states. In contemporary global justice debates, there is deep disagreement about the extent of such redistributive claims. On the one hand, cosmopolitans argue that robust principles of distributive justice ought to apply globally, while minimalists argue that stronger principles apply within states. This division has been profoundly influenced by John Rawls's *A Theory of Justice* (1971), and in particular his second principle of justice. This actually comprises two sub-principles, namely a "fair equality of opportunity principle," and a "difference principle" requiring any remaining inequalities among individuals to benefit the worst off in society. Cosmopolitan interpreters argue that both principles ought to apply globally (Beitz, 1975; Pogge, 1989, 2002; Moellendorf, 2002; Caney, 2005). However, Rawls himself argued that these principles apply only within states, recognizing only weaker global principles (Rawls 1999). Depending upon one's conception of justice, one will recognize more or less stringent redistributive demands.

Many theories would also differentiate between normative claims to revenues drawn from presently unowned resources, such as carbon emissions, and claims to revenues from land or natural resources, which are variously owned, taxed, and traded already. Carbon rents are an illustrative case. Most accounts of climate justice conceive of the atmosphere as an unowned common, and proceed in distributing rights to emit on the basis of basic equality of entitlement (Singer, 2016; Vanderheiden, 2008). Given the absence of plausible claims to the atmosphere, carbon rents might be subjected to various principles of redistribution, including equality, according to which all benefit equally; sufficiency, according to which benefits bring those in deprivation to a normatively adequate level of well-being; or priority, according to which benefits go to the very worst off. But because carbon pricing regimes envisage rents from terrestrial carbon sinks, for instance through the UN's REDD+ program, global redistributive claims might be resisted. This is not simply because the doctrine of permanent sovereignty over natural resources is recognized in international law, since this fact has no obvious normative significance for distributive justice (Armstrong, 2015). Instead, claims to terrestrial carbon sinks may be differentiated from rents from carbon emissions on the basis of local entitlements or attachments to land and resources (Blomfield, 2013). All the same, it is not clear that existing normative claims based upon improvement or attachment can support claims to benefit from carbon storage, which has only recently become a potentially valued resource (Armstrong, 2017).

Although there is no universally accepted perspective of what constitutes an equitable sharing of the climate rent, the above considerations can guide an arrangement that would be widely accepted. A globally just distribution of the climate rent is closely tied to the promotion of well-being within individual countries, which we analyze in the Section 3.2.4.

3.2.4. *Using Carbon Pricing Revenues to Promote Human Development*

The "double dividend" literature has examined how fiscal policies to address environmental externalities, such as emissions pricing, interact with the overall tax system. This literature has emphasized the efficiency gains that can be achieved by lowering other distortionary taxes (Goulder, 2013). In contrast, relatively few studies have assessed the potential to use the revenues from emission pricing to increase public spending, which could advance social justice by improving the situation of the poorest members of society in line with the SDGs. This could be of particular importance for developing countries, which often face constrained fiscal space. From this perspective, the central objective of climate policy consists in advancing human needs and human well-being (Gough, 2015; Lamb and Steinberger, 2017).

Drawing on the median revenue projections of the 2°C scenarios shown in Figure 3.1, Jakob et al. (2016) calculate the extent to which carbon pricing revenues could finance access to basic infrastructure services. Their benchmark scenario assumes that no redistribution across countries takes place (i.e. the globally optimal carbon price is applied in each country, without side payments or trade of emission payments). In this case, they find that for all regions except sub-Saharan Africa (which at the same time displays the largest access gaps and lowest emissions, and hence the lowest revenues), carbon pricing revenues would be sufficient to finance universal access to water, sanitation, or electricity by 2030. In an alternative scenario, in which distribution of revenues is the average between the above scheme and an equal per capita allocation of emission rights, sub-Saharan Africa receives a substantially higher share of the global revenues, such that all countries would be able to achieve universal access to water, sanitation, or electricity. In a similar vein, Jakob et al. (2015) demonstrate that removing existing consumption subsidies for fossil fuels and redirecting the freed financial resources to infrastructure investment would allow about seventy countries to achieve universal water access, more than sixty countries would have universal access to sanitation, and more than fifty countries could provide access to electricity for their entire population by 2030. Finally, Franks et al. (n.d.) consider additional investments related to achieving the SDGs, such as transport, health, education, and food security. Relying on the 2°C scenario from Figure 3.1 and assuming a complete phase-out of fossil fuel subsidies, they find that very few countries would be able to *fully* finance SDG investment needs from the additional revenues from such fiscal reforms. However, a substantial amount of countries, especially in South and South East Asia, could meet *more than half* of their financing needs from this novel source of public revenues.

3.3. Resource and Land Rents

Institutions such as international agreements create scarcity for the use of the atmosphere as a disposal space for GHGs. In contrast, land and natural resource rents arise from their natural scarcity. Even though some have argued in favor of a global governance regime for land (Creutzig, 2017), ownership of land and natural resources is in most instances confined to national boundaries—another marked difference from the atmosphere. This section discusses to what extent land and resource rents can contribute to public finances and examines equity aspects relevant for their just distribution.

3.3.1. *Taxing Land and Natural Resource Rents*

Natural resource rents are determined by the difference between the price of a commodity and extraction costs. The World Bank (2011) estimated that in 2010, global natural resource rents amounted to more than $3 trillion, or slightly less than 5 percent of global GDP. These resources include forests, oil, gas, coal, and minerals (bauxite, copper, lead, nickel, phosphate, tin, zinc, gold, silver, and iron), of which fossil fuels account for the lion's share. Estimates for land rents show substantial variation, depending on what kinds of land (e.g. agricultural vs. urban) are included in the analysis and the valuation of land-based amenities that are usually not traded on markets. Based on household data from selected developing countries in which respondents were asked to assess the economic value of the land they own or occupy, Kalkuhl et al. (n.d.) find land rents ranging from about 1 percent to roughly 7.5 percent of GDP.

Land rent taxation is a central issue in classical economics (George, 1879). Modern economic theory has reaffirmed that taxing fixed factors, such as land and natural resources, could increase economic efficiency (Norregaard, 2013). In contrast, most taxes that are currently applied (e.g. on labor or capital) induce economic distortions. They may do this by reducing the incentive to work or save (Feldstein, 1999). By contrast, taxes on fixed factors are either neutral (i.e. non-distortionary) or could even increase economic efficiency by addressing underinvestment in capital. Such an underinvestment may occur as a result of an oversized propensity to use savings to acquire fixed assets, such as land or natural resources (Edenhofer, Mattauch, and Siegmeier, 2015). Land rent taxation could also be beneficial for the conservation of forests and ecosystems by reducing the incentive to convert areas to agricultural land (Kalkuhl and Edenhofer, 2017), and for the reduction of urban sprawl by encouraging higher density in urban areas (Banzhaf and Lavery 2010).

For the aforementioned reasons, land rent taxes are promising tools for domestic resource mobilization. Kalkuhl et al. (n.d.) carry out micro-simulations to assess the possible effects of land rent taxation for a number of selected countries. They find that taxing half of these rents would increase public revenues on average by about 15 percent. However, since in many countries poor households hold a substantial share of their wealth in the form of land, such an approach would have regressive effects on income distribution. In other words, measured as a share of disposable income, land taxes would put a higher cost burden on poorer households (see Figure 3.2). This outcome can be avoided by introducing an allowance that is not subject

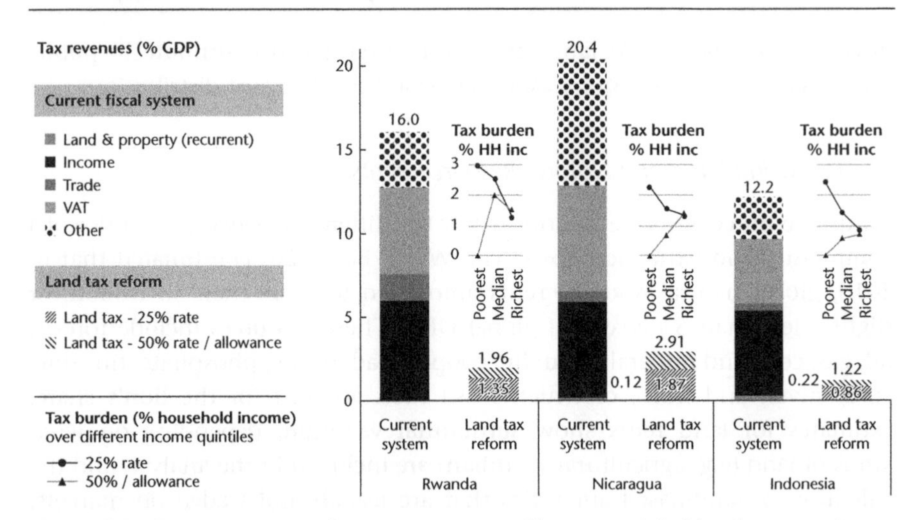

Figure 3.2. Current sources of tax revenues as well as potential of land taxes and distributional effects in Rwanda, Nicaragua, and Indonesia
Source: Kalkuhl et al. (2017).

to taxation. Such a provision would reduce public revenue by about one-third, but might be preferable in terms of social justice and political feasibility.

3.3.2. *Justice Arguments Concerning Natural Resource and Land Rents*

There are long-running debates between cosmopolitans and their opponents about the just global distribution of resource revenues. Prominent positions thus have clear implications for how a new revenue stream might be justly distributed. Because some cosmopolitans believe that Rawls's difference principle ought to apply globally, any inequalities in the distribution of natural resources would be tolerated only if they benefited the worst off (e.g. Beitz, 1975). Pogge (2002) influentially argued that the global order fails in this regard, by systematically undermining human rights. For this reason, Pogge argues for basic resource entitlements for all to live minimally decent lives. Capabilities theorists would similarly require greater redistribution of resource values if their underprovision systematically threatened the ability of some to live minimally decent lives (Nussbaum and Sen, 1993; Sen, 2009). More radical still are egalitarian and utilitarian theories. Armstrong's (2017) egalitarian theory holds that all important sources of advantage and disadvantage ought to be distributed according to a principle of equality of opportunity, while utilitarians require the allocation of resources to maximize global well-being or happiness (Singer, 2016). Instead, non-cosmopolitan minimalists generally reject global redistribution, recognizing more robust obligations to fellow citizens or members of a nation. Risse (2012) accepts Rawls's restriction of the difference principle to individual societies, although he recognizes a minimal common ownership right to merely use natural resources in order to satisfy basic human needs. Miller (2007) and Moore (2001) both argue for the normative particularity of the nation as an ethical community or relationship, which involves justifying claims to resource values in order to be self-determining. Nonetheless, minimalists could hold that nations enjoy claims to revenues from natural resources, while granting that some proportion of these ought to be redistributed in the interests of extreme deprivation elsewhere (Moore, 2012).

Global justice debates often focus upon resources such as oil or diamonds that appear little affected by human endeavor. Such resources are thus subject to either improvement or attachment-based special claims. Both improvement and attachment-based claims may be most plausible for land. This implies that any potential redistribution of land rents would be highly contentious. Although some follow Henry George in proposing land value taxation for efficiency reasons (Stiglitz, 2016), it is an open question whether this could be supported by considerations of justice. One attempt is made by Steiner (1999), who argues for redistribution of all "unearned" benefits, such as increases in land and natural resource values. However, this proposal is vulnerable to the objection that there is no way to disentangle natural from culturally produced values (Miller, 2007, pp. 59–60). Nonetheless, egalitarians might support redistribution of a proportion of land value once plausible improvement or attachment-based claims have been satisfied.

While theories differ concerning the robustness of global redistribution, and concerning the normative status of particular resources, redistributive claims may be strengthened by the shared global commitment to achieve the SDGs. Given the great inequalities that persist among nation states, and the relative ease with which some can achieve these goals, redistribution of at least a proportion of resource revenues appears plausible according to most theories of justice, even if the strength of such claims can be expected to vary. In the case of climate rents, the justification for redistribution is bolstered by evidence concerning the ability of rents to reduce severe deprivation, in line with the SDG agenda.

3.3.3. *Using Land and Resource Rents to Promote Human Development*

Segal (2010) highlights that purely domestic redistribution of resource rents on an equal per capita basis could reduce the number of people living on less than $1 a day globally by up to two-thirds. Resource rent taxation also constitutes a promising option to finance basic infrastructure, at least as part of the required investments to close these gaps. From this perspective of investment needs, Fuss et al. (2016) calculate that using rents accruing from natural resources on the domestic level (i.e. without any kind of redistributions across countries) would provide access to water, sanitation, electricity, and telecommunications for the majority of countries. Figure 3.3 displays the share of resource rents that would be needed to close existing access gaps by 2030. In fact, many resource-rich countries could meet their financing needs in each of these areas by using less than 10 percent of their resource rents.

Some authors have claimed that natural resource rents accrue first and foremost because of market power rather than from resource scarcity (Hart and Spiro 2011). For instance, Hamilton (2009) argues that "scarcity rent made a negligible contribution to the price of oil." Without judging the validity of this argument, we would like to examine its implications for rent taxation. If it is true that rents to resource owners can be characterized as monopoly rents, then the economically optimal solution mandates addressing this market power, e.g. by means of regulation. However, as markets for natural resources are global in scope, such an approach would require international cooperation. This might be difficult to achieve, especially for relatively poor countries possessing little influence on the formulation of global policies. As a consequence, a robust second-best policy may be to tax the

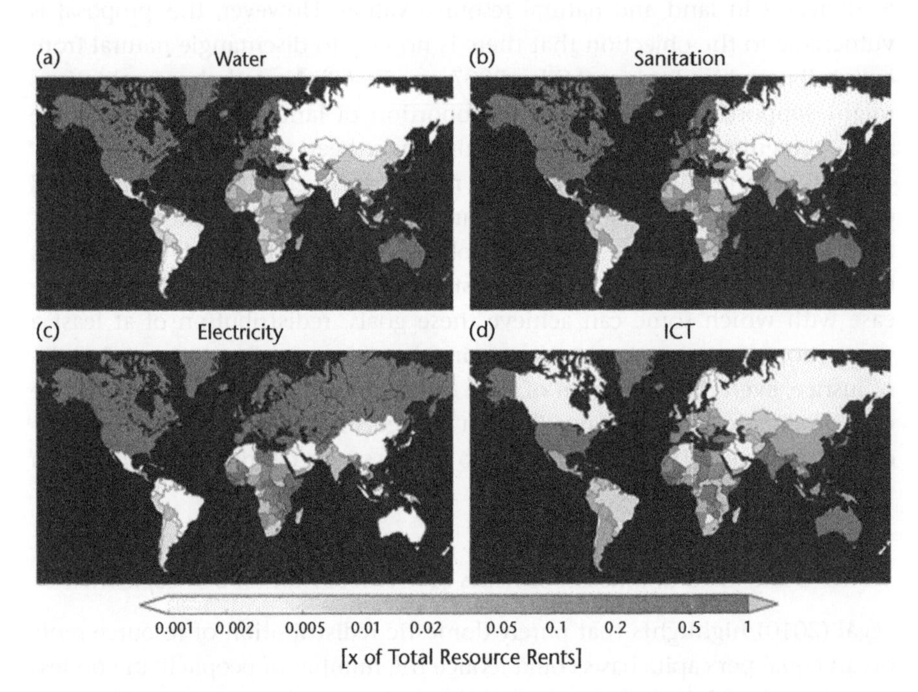

Figure 3.3. Total share of resource rents needed to simultaneously achieve universal access to (a) water, (b) sanitation, (c) electricity, and (d) telecommunication

Note: logarithmic scale.

Source: Fuss et al. (2016).

profits accruing to firms that are extracting and dealing with natural resources. Such a policy applied in cases where underlying market distortion cannot be readily removed would transfer monopoly rents into public revenues.

3.4. Political Feasibility and Implementation

Although internalizing externalities and taxing rents can be expected to raise social welfare, the political feasibility of such options depends on the use of the associated public revenues. This section addresses the question of how the attractiveness of such approaches can be increased for policymakers, and examines issues associated with concrete implementation.

3.4.1. *Political Feasibility*

Carbon pricing can be expected to entail costs for individual households, e.g. in the form of rising electricity bills, and the costs of heating and transportation, as well as higher prices of goods and services that use energy as an input in production. In high-income countries, poorer households frequently spend a higher share of their income on energy-intensive goods and services. Carbon prices might then have a regressive effect, i.e. imposing a higher cost as a share of income on poorer households. Performing micro-simulations based on household survey data for roughly eighty countries, Dorband et al. (n.d.) demonstrate that for the majority of countries, carbon pricing would have a progressive impact on the distribution of income. That is, carbon pricing tends to be progressive for countries with per capita GDP below $8,000, but turns regressive at higher income levels. However, even in cases in which progressive results are obtained, carbon pricing could still entail sizable costs for the poorest segment of the population.

Several options have been proposed to avoid adverse impacts on the poor. In theory, revenues could in most cases be redistributed in a way that makes taxation progressive. In reality, however, discretionary spending is usually impossible and would be hampered by information and transaction costs. Hence, the chosen revenue recycling scheme is crucially important. Popular designs being discussed include per capita recycling (e.g. in form of a so-called "tax-and-dividend" scheme) and targeted reductions of taxes and levies that would favor low-income households (Klenert et al., 2017). Recent carbon pricing schemes have, to at least some extent, adopted these approaches. For instance, British Columbia has designed its carbon tax in a revenue-neutral way, in which about 40 percent is reimbursed to households by income tax cuts as well as lump-sum transfers (Beck et al., 2015).

For developing countries, political feasibility may increase if the revenues resulting from a price on a commons are targeted toward investments that benefit low-income households. For instance, in Nigeria, Dorband (2016) find that carbon pricing would impose the proportionally lowest costs on poor households. At the same time, these households are most affected by under-provision of basic infrastructure and would hence gain the most from increased provision. For this reason, combining carbon pricing with dedicated infrastructure spending would result in "double progressivity."

Another area in which revenues will need to be employed concerns the compensation of powerful lobby groups, which might otherwise resist the introduction of such a policy (Trebilcock, 2014). A recent study based on integrated climate-energy-economy modeling (Bauer et al., 2013) estimates that limiting the atmospheric concentration of CO_2 to a level that would be consistent with the 2°C target (450 parts per million (ppm)) would reduce the net present value of fossil fuels by about $12 trillion (see Figure 3.4). Even if this target were to be relaxed to 550 ppm, which would correspond to a temperature increase of roughly 3°C, declining demand for fossil fuels would still reduce resource owners' rents by more than $8 trillion. Fossil fuel owners thereby have substantial incentives to oppose climate policies.

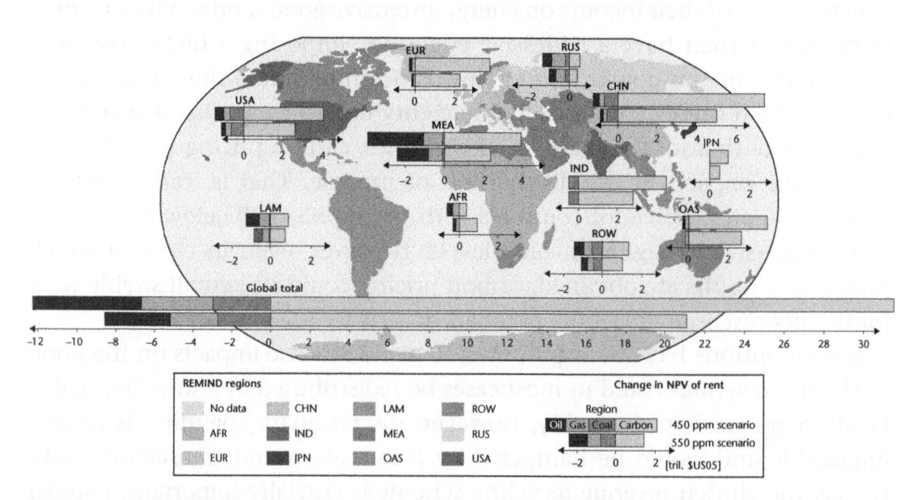

Figure 3.4. Projected changes in economic rents accruing to fossil energy carriers and carbon in mitigation scenarios, relative to baseline, over the twenty-first century (in US$ trillion)

The upper bar indicates the 450 ppm scenario, the lower bar the 550 ppm scenario. Both scenarios assume full availability of mitigation technologies and introduction of a global carbon price without delay. Globally, declines of rents from oil (left bar, leftmost segment), gas (left bar, middle segment), and coal (left bar, rightmost segment) resulting from climate policy are overcompensated by the increasing carbon rent (right bar).

Source: Data from Bauer et al. (2013).

On aggregate, however, such losses would be overcompensated by the rents that are created by climate measures. For the 450 ppm scenario, this would amount to more than $30 trillion, and for the 550 ppm scenario to more than $20 trillion. In regions such as Latin America and the Caribbean (LAM) as well as the Middle East and North Africa (MEA), losses are comparable or even higher than the value of the climate rent generated by carbon pricing. As such, these regions would likely require some kind of compensation in addition to just sharing mitigation costs, e.g. in terms of financial payments or technology transfer.

On the country level, the climate rent could be collected by, *inter alia*, a carbon tax or auctioning of emission permits. That rent could be used to increase the political acceptance of climate measures, for example by compensating fossil fuel owners for some of their losses. In emission trading schemes, energy-intensive firms and utilities have frequently been compensated by the free allocation of emission permits, which has resulted in large windfall profits. Yet, awarding emission permits on the basis of past emissions (as, e.g., under a "grandfathering" scheme) creates a perverse incentive to "ratchet up" emissions in order to obtain more free allocations in the future (Weitzman, 1980). It needs to be ensured that these free allocations cannot be carried forward indefinitely—they must include sunset clauses. Even more importantly, capacity building (e.g. scenario studies also industry sectors) and R&D investments are required to prepare and enable future decarbonization efforts.

3.4.2. *Implementation*

Putting a price on a common and using the revenues to promote social objectives can be regarded as a clear case of "earmarking," which is a highly contentious issue. On the one hand it is seen to reduce flexibility and incentivize misspending. Others have argued that dedicating the revenues from an environmental tax to a specific issue that is seen to be socially desirable can greatly increase social support (Kallbekken et al., 2011).

Introducing carbon pricing can be expected to face important institutional and political barriers, such as governments lacking the ability to credibly commit to long-term policy (Nemet et al., 2017). For this reason, feasible policies need to be tailored to the specific economic, political, and institutional context in which they operate. This might also explain why, in the real world, combinations of several policies are frequently observed. According to economic theory, the economically optimal solution consists in choosing one policy instrument for each market failure. It has been argued that for the case of climate policy, technology policies are required to complement emission prices as they internalize positive externalities through learning-by-doing

(Jaffe et al., 2005). Yet, most countries display combinations of different policies that contradict this rule, such as emissions pricing combined with efficiency standards. These policy combinations can be understood from a second-best perspective to either alleviate distributional concerns if more efficient mechanisms (such as lump-sum payments) are not available to policymakers, or to increase credibility that the overall policy framework remains in place even if an individual instrument is revoked (Nemet et al., 2017).

Such "second-best" considerations play an important role in dynamic settings, in which policies can be introduced in a sequence to build up "winning coalitions" that can be expected to support more ambitious policies in the future. For instance, Meckling et al. (2015) argue that technology policies constitute an essential foundation for the introduction of carbon prices, as technological progress lowers the carbon price required to achieve a given emission target, and hence eases distributional struggles. For the case of Ecuador, Jakob (2017) provides examples of policies that could be politically feasible to implement and that might prepare the ground for a reform of fossil fuel subsidies. These include the reform of driving restrictions in urban areas, expansion of public transport, and financial support for electric vehicles, which would make rising prices for transport fuels more acceptable by dampening their adverse impacts on household incomes.

3.5. Conclusions

We have argued that the atmosphere currently constitutes a common pool resource. Its sustainable management in the form of a global common property regime requires international cooperation and coordination among self-interested nation states. International cooperation could be strengthened by a combination of carbon pricing and conditional transfers. The revenues from carbon pricing and the taxation of land and resource rents could aid governments in their efforts to promote human development targets. We have used the SDG agenda to motivate an understanding of human well-being from a multidimensional perspective. Avoiding degradation of natural capital, the reduction of inequality and access to essential infrastructure services are at the core of this analysis. Using the revenues from taxing GHGs as well as land and resource rents to advance the SDG agenda could help to alleviate the trade-offs between economic efficiency and social justice.

Political feasibility is likely to be the most important obstacle for sustainable management of the commons and use of revenues in a way that promotes human development. Lack of credible commitments and powerful vested interests have the potential to undermine the trust in government, which is essential for citizens to accept taxation.

The approach to use rent taxation to further human development targets put forward in this chapter is less ambitious than many cosmopolitan or utilitarian theories of justice because the revenues under consideration are primarily within national boundaries. However, it is more ambitious than theories of minimalistic national justice because it argues for robust international support to enable poor countries to participate effectively in international agreements. While the approach proposed in this chapter is admittedly demanding, it might nonetheless be feasible and may even be preferable compared to more demanding proposals requiring sweeping global governance reform.

Acknowledgments

We thank Matthias Kalkuhl for providing the data for Figure 3.2, Nico Bauer for providing the data for Figure 3.4, Martin Wodinski for designing Figure 3.4, and Kristin Seyboth for language editing.

References

Armstrong, Chris. 2015. "Against 'Permanent Sovereignty' Over Natural Resources." *Politics, Philosophy & Economics* 14 (2): 129–51.

Armstrong, Chris. 2017. *Justice and Natural Resources: An Egalitarian Theory*. Oxford: Oxford University Press.

Banzhaf, H. Spencer, and Nathan Lavery. 2010. "Can the Land Tax Help Curb Urban Sprawl? Evidence from Growth Patterns in Pennsylvania." *Journal of Urban Economics* 67 (2): 169–79.

Barrett, Scott. 2005. *Environment and Statecraft: The Strategy of Environmental Treaty-Making*. Oxford: Oxford University Press.

Bauer, Nico, Ioanna Mouratiadou, Gunnar Luderer, Lavinia Baumstark, Robert J. Brecha, Ottmar Edenhofer, and Elmar Kriegler. 2013. "Global Fossil Energy Markets and Climate Change Mitigation—an Analysis with REMIND." *Climatic Change* 136 (1): 1–14.

Beck, Marisa, Nicholas Rivers, Randall Wigle, and Hidemichi Yonezawa. 2015. "Carbon Tax and Revenue Recycling: Impacts on Households in British Columbia." *Resource and Energy Economics* 41 (August): 40–69.

Beitz, Charles R. 1975. "Justice and International Relations." *Philosophy & Public Affairs* 4 (4): 360–89.

Blanford, Geoffrey J., Elmar Kriegler, and Massimo Tavoni. 2014. "Harmonization vs. Fragmentation: Overview of Climate Policy Scenarios in EMF27." *Climatic Change* 123 (3–4): 383–96.

Blomfield, Megan. 2013. "Global Common Resources and the Just Distribution of Emission Shares." *Journal of Political Philosophy* 21 (3): 283–304.

Caney, Simon. 2005. *Justice Beyond Borders*. Oxford: Oxford University Press.

Carbon Pricing Leadership Coalition. 2017. "Report of the High-Level Commission on Carbon Prices." <https://www.carbonpricingleadership.org/report-of-the-highlevel-commission-on-carbon-prices/>.

Cramton, Peter, Axel Ockenfels, and Steven Stoft. 2015. "An International Carbon-Price Commitment Promotes Cooperation." *Economics of Energy & Environmental Policy* 4 (2): 51–64.

Creutzig, Felix. 2017. "Govern Land as a Global Commons." *Nature* 546 (7656): 28–9.

Dorband, Ira. 2016. "Using Revenues from Carbon Pricing to Close Infrastructure Access Gaps—Distributional Impacts on Nigerian Households." Master thesis. FU Berlin.

Dorband, Ira, Michael Jakob, Matthias Kalkuhl, and Jan Steckel. n.d. "Are Poor Households More Strongly Impacted by Carbon Pricing? A Global Comparative Analysis of Distributional Effects."

Edenhofer, Ottmar, Christian Flachsland, and Ulrike Kornek. 2016. "Der Grundriss Für Ein Neues Klimaregime." *Ifo Schnelldienst* 69 (3): 11–15.

Edenhofer, Ottmar, Michael Jakob, Felix Creutzig, Christian Flachsland, Sabine Fuss, Martin Kowarsch, Kai Lessmann, Linus Mattauch, Jan Siegmeier, and Jan Christoph Steckel. 2015. "Closing the Emission Price Gap." *Global Environmental Change* 31: 132–43.

Edenhofer, Ottmar, Linus Mattauch, and Jan Siegmeier. 2015. "Hypergeorgism: When Rent Taxation Is Socially Optimal." *FinanzArchiv: Public Finance Analysis* 71 (4): 474–505.

Feldstein, Martin. 1999. "Tax Avoidance and the Deadweight Loss of the Income Tax." *Review of Economics and Statistics* 81 (4): 674–80.

Franks, Max, Kai Lessmann, Ottmar Edenhofer, Michael Jakob, and Jan Steckel. forthcoming "Carbon Pricing Could Mobilize Domestic Resources for the Agenda 2030." *Nature Sustainability*

Fuss, Sabine, Claudine Chen, Michael Jakob, Annika Marxen, Narasimha D. Rao, and Ottmar Edenhofer. 2016. "Could Resource Rents Finance Universal Access to Infrastructure? A First Exploration of Needs and Rents." *Environment and Development Economics* 21 (6): 691–712.

Gardiner, Stephen M., Simon Caney, Dale Jamieson, Henry Shue, and Rajendra Kumar Pachauri. 2010. *Climate Ethics: Essential Readings*. Oxford: Oxford University Press.

George, Henry. 1879. *Progress and Poverty*. New York: Robert Schalkenbach Foundation.

Gough, Ian. 2015. "Climate Change and Sustainable Welfare: The Centrality of Human Needs: Fig. 1." *Cambridge Journal of Economics* 39 (5): 1191–214.

Goulder, Lawrence H. 2013. "Climate Change Policy's Interactions with the Tax System." *Energy Economics* 40 (December): S3–11.

Hamilton, James D. 2009. "Understanding Crude Oil Prices." *The Energy Journal* 30 (2): 179–206.

Hart, Rob, and Daniel Spiro. 2011. "The Elephant in Hotelling's Room." *Energy Policy* 39 (12): 7834–38.

IIASA. 2014. *IPCC AR5 Scenarios Database*. <http://www.iiasa.ac.at/web/home/research/researchPrograms/Energy/IPCC_AR5_Database.html>.

Jaffe, Adam B., Richard G. Newell, and Robert N. Stavins. 2005. "A Tale of Two Market Failures: Technology and Environmental Policy." *Ecological Economics* 54 (2–3): 164–74.

Jakob, Michael. 2017. "Ecuador's Climate Targets: A Credible Entry Point to a Low-Carbon Economy?" *Energy for Sustainable Development* 39 (August): 91–100.

Jakob, Michael, Claudine Chen, Sabine Fuss, Annika Marxen, and Ottmar Edenhofer. 2015. "Development Incentives for Fossil Fuel Subsidy Reform." *Nature Climate Change* 5 (8): 709–12.

Jakob, Michael, Claudine Chen, Sabine Fuss, Annika Marxen, Narasimha D. Rao, and Ottmar Edenhofer. 2016. "Carbon Pricing Revenues Could Close Infrastructure Access Gaps." *World Development* 84: 254–65.

Jakob, Michael, and Ottmar Edenhofer. 2014. "Green Growth, Degrowth, and the Commons." *Oxford Review of Economic Policy* 30(3): 447–68.

Jakob, Michael, and Jan Christoph Steckel. 2014. "How Climate Change Mitigation Could Harm Development in Poor Countries." *Wiley Interdisciplinary Reviews: Climate Change* 5 (2): 161–68.

Kalkuhl, Matthias, and Ottmar Edenhofer. 2017. "Ramsey Meets Thünen: The Impact of Land Taxes on Economic Development and Land Conservation." *International Tax and Public Finance* 24 (2): 350–80.

Kalkuhl, Matthias, Blanca Fernandez Milan, Gregor Schwerhoff, Michael Jakob, Maren Hahnen, and Felix Creutzig. forthcoming "Fiscal Instruments for Sustainable Development: The Case of Land Taxes." *Land Use Policy*.

Kalkuhl, Matthias, Blanca Fernandez Milan, Gregor Schwerhoff, Michael Jakob, Maren Hahnen, Felix Creutzig, Jetske Bouma, and Stefan van der Esch. 2017. "Land Taxes as Fiscal Instruments to Promote Sustainable Development." <http://www.pbl.nl/sites/default/files/cms/publicaties/MCC-PBL-2017-Fiscal-instruments-for-sustainable-development-the-case-of-land-taxes-2735.pdf>.

Kallbekken, Steffen, Stephan Kroll, and Todd L. Cherry. 2011. "Do You Not Like Pigou, or Do You Not Understand Him? Tax Aversion and Revenue Recycling in the Lab." *Journal of Environmental Economics and Management* 62 (1): 53–64.

Klenert, David, Linus Mattauch, Emmanuel Combet, Ottmar Edenhofer, Cameron Hepburn, Ryan Rafaty, and Nicholas Stern. 2017. "Making Carbon Pricing Work." MPRA Paper No. 80943. <https://mpra.ub.uni-muenchen.de/80943/1/MPRA_paper_80943.pdf>.

Kornek, Ulrike, and Ottmar Edenhofer. 2016. "The Strategic Dimension of Financing Global Public Goods." EAERE Conference paper.

Kossoy, Alexandre, Grzegorz Peszko, Klaus Oppermann, Nicolai Prytz, Noémie Klein, Kornelis Blok, Long Lam, Lindee Wong, and Bram Borkent. 2015. *State and Trends of Carbon Pricing 2015*. Washington, DC: World Bank.

Kriegler, Elmar, John P. Weyant, Geoffrey J. Blanford, Volker Krey, Leon Clarke, Jae Edmonds, Allen Fawcett, et al. 2014. "The Role of Technology for Achieving Climate Policy Objectives: Overview of the EMF 27 Study on Global Technology and Climate Policy Strategies." *Climate Change* 123: 353–67.

Lamb, William F., and Julia K. Steinberger. 2017. "Human Well-Being and Climate Change Mitigation: Human Well-Being and Climate Change Mitigation." *Wiley Interdisciplinary Reviews: Climate Change* 8 (6): e485.

Lessmann, Kai, Ulrike Kornek, Valentina Bosetti, Rob Dellink, Johannes Emmerling, Johan Eyckmans, Miyuki Nagashima, Hans-Peter Weikard, and Zili Yang. 2015. "The

Stability and Effectiveness of Climate Coalitions: A Comparative Analysis of Multiple Integrated Assessment Models." *Environmental and Resource Economics* 62 (4): 811–36.

MacKay, David J. C., Peter Cramton, Axel Ockenfels, and Steven Stoft. 2015. "Price Carbon—I Will if You Will." *Nature* 526 (7573): 315–16.

Markandya, Anil. 2011. "Equity and Distributional Implications of Climate Change." *World Development* 39 (6): 1051–60.

Meckling, J., N. Kelsey, E. Biber, and J. Zysman. 2015. "Winning Coalitions for Climate Policy." *Science* 349 (6253): 1170–1.

Miller, David. 2007. *National Responsibility and Global Justice*. Oxford: Oxford University Press.

Moellendorf, Darrel. 2002. *Cosmopolitan Justice*. Boulder, CO: Westview Press.

Moore, Margaret. 2001. *The Ethics of Nationalism*. Oxford: Oxford University Press.

Moore, Margaret. 2012. "Natural Resources, Territorial Right, and Global Distributive Justice." *Political Theory* 40 (1): 84–107.

Nemet, Gregory F., Michael Jakob, Jan Christoph Steckel, and Ottmar Edenhofer. 2017. "Addressing Policy Credibility Problems for Low-Carbon Investment." *Global Environmental Change* 42 (January): 47–57.

Norregaard, John. 2013. "Taxing Immovable Property Revenue Potential and Implementation Challenges." International Monetary Fund.

Nussbaum, Martha C., and Amartya Sen, eds. 1993. *The Quality of Life*. Oxford: Oxford University Press.

Pogge, Thomas. 1989. *Realizing Rawls*. Ithaca, NY: Cornell University Press.

Pogge, Thomas. 2002. *World Poverty and Human Rights*. Cambridge: Polity Press.

Rawls, John. 1971. *A Theory of Justice*. Cambridge, MA: Harvard University Press.

Rawls, John. 1999. *The Law of Peoples, with the Idea of Public Reason Revisited*. Cambridge, MA: Harvard University Press.

Risse, Matthias. 2012. *On Global Justice*. Princeton, NJ: Princeton University Press.

Segal, Paul. 2010. "Resource Rents, Redistribution, and Halving Global Poverty: The Resource Dividend." *World Development* 39 (4): 475–89.

Sen, Amartya. 2009. *The Idea of Justice*. London: Penguin Books.

Singer, Peter. 2016. *One World Now: The Ethics of Globalization*. Revised 3rd edition. Terry Lecture Series. New Haven, CT: Yale University Press.

Steiner, Hillel. 1999. "Just Taxation and International Redistribution." *Nomos* 41: 171–91.

Stiglitz, Joseph E. 2016. "How to Restore Equitable and Sustainable Economic Growth in the United States." *American Economic Review* 106 (5): 43–7.

Trebilcock, Michael J. 2014. *Dealing with Losers: The Political Economy of Policy Transitions*. Oxford: Oxford University Press.

UNFCCC. 2015. "Draft Decision -/CP.21. Adoption of the Paris Agreement." <http://unfccc.int/resource/docs/2015/cop21/eng/l09r01.pdf>.

Vanderheiden, Steve. 2008. *Atmospheric Justice: A Political Theory of Climate Change*. Oxford and New York: Oxford University Press.

Weitzman, Martin L. 1980. "The 'Ratchet Principle' and Performance Incentives." *Bell Journal of Economics* 11 (1): 302–8.

World Bank. 2011. "The Changing Wealth of Nations." <http://siteresources.worldbank.org/ENVIRONMENT/Resources/ChangingWealthNations.pdf>.

4

Equity Implications of the COP21 Intended Nationally Determined Contributions to Reduce Greenhouse Gas Emissions

Adam Rose, Dan Wei, and Antonio Bento

4.1. Introduction

In 2015, at the twenty-first Conference of the Parties to the UN Framework Convention on Climate Change (COP21), 195 countries made pledges to reduce greenhouse gas (GHG) emissions to avoid a 2°C rise in global average temperatures. This was an unprecedented convergence of international resolve to combat climate change. It came after more than twenty-five years of negotiations to reach a truly global accord on this pressing issue. During that period, one of the major obstacles to reaching agreement was how to share the cost burden of GHG mitigation and sequestration. Equity, or fairness, is at the core of the issue. Many equity principles were explicitly or implicitly invoked, and negotiations proceeded by attempting to reach an agreement among countries on a single or combination of burden-sharing principles. However, this "top-down" approach failed to achieve its goals.[1]

[1] Some have argued that top-down equity principles have been more of a focus of academic discourse than actual policymaking. Several countries proposed equity principles of their own without regard to the decisions of others, while others proposed emissions reduction commitments without reference to equity principles at all. However, country positions in the course of negotiations on climate agreements have in fact revolved around equity explicitly and implicitly; most notably, using different equity arguments, the US never joined the Kyoto Protocol, and China, as well as the majority of developing countries, made no emissions reduction commitment under the Protocol. Some also point to the Kyoto Protocol commitments being couched in terms of "common but differentiated responsibilities" as an indication that signatories were able to move past the equity principles stifling progress, but Rose et al. (1998) and others have made a strong case that the pleading of special considerations by signatory

Momentum started to build over the years behind an alternate approach, termed "bottom-up," by which a more decentralized decision-making process, as well as a broader set of ancillary goals, could achieve global agreement (Rayner, 2010; Andresen, 2015). COP21 is an ultimate example, where each signatory offered its own pledge to reduce GHGs, based on its own criteria, including, for many, fairness. Although countries were aware of each other's pledges, they seemed to have little explicit concern about the pledges of others in specifying their own.

COP21 is clearly a victory for the bottom-up approach and a major step forward in climate negotiations. However, it represents a substantial change in the positions of many developing countries on the international equity of burden sharing. For example, Mexico, as did many developing countries, favored the Egalitarian equity approach for many years. Applying this principle, Mexico's GHG emission reduction percentage in 2030 would have been near zero, but it pledged a 25 percent reduction. GHG percentage reduction pledges for 2030 for most developing countries appear similarly high given expectations of strong economic growth and hence emissions growth. Such changes in the positions of these developing countries raise the question of whether the pledges made at the COP21 are their final positions, and, if they are, what are the prospects of these pledges being carried out? Another concern is whether the promised financial transfers of at least $100 billion per year from industrialized to developing countries to support climate mitigation initiatives in the latter countries will be forthcoming. Averchenkova and Bassi (2016), for example, have concluded that GHG reduction pledges similar to those at Paris for a sample of countries at various stages of development are only moderately credible.

The purpose of this chapter is to analyze the equity implications of the bottom-up approach to climate change negotiations in two major ways. First, we analyze the GHG reduction pledges specified in the recent COP21 agreement prior to any emissions trading through the adaptations of the equity metrics of the Gini coefficient and Atkinson index. We also compare the distribution of the pledges with specific equity principles, such as the Egalitarian, Ability to Pay, Vertical Equity, Horizontal Equity, and Rawlsian Maximin principles (Kverndokk and Rose, 2008; IPCC 2014).

Second, we analyze the equity outcomes after emissions allowance trading takes place by adapting earlier modeling efforts by the authors (Rose et al., 1998; Rose and Wei, 2008). Emissions trading is a promising approach to lowering the overall cost of meeting the Paris targets and for providing developing countries with financing by selling allowances. An innovative aspect

countries, such as Australia (high dependence on coal exports) and Russia (transitioning economy), for example, were merely an appeal to fairness in terms of economic hardships.

of this analysis is that it includes sensitivity tests on macroeconomic vs. microeconomic costs, an alternative future technology, and the use of offsets.

Our quantitative analysis of the effect of emissions allowance trading on equity outcomes focuses on three representative regions/countries—China, the European Union, and California—but we also discuss the implications for additional countries. Our choice of countries/regions is related to recent developments of non-standard approaches to climate change negotiations, as well as to limitations of available data. For example, California itself is the sixth largest economy in the world, and its current and former governors have been at the forefront of climate change policy initiatives. On one of his last official activities, UK Prime Minister Tony Blair visited California to discuss the prospects of emissions trading between the state and the EU. The Trump administration's decision to pull out of the Paris Agreement makes the role of California and other US states and regions in achieving global GHG emission reduction targets all the more important.[2] The analytical framework is, of course, generalizable to the 195 counties that are parties to the COP21 Accord.

Our analysis indicates that the GHG reduction pledges made in the Paris Agreement reflect substantial inequality based on the evaluation of some standard equity metrics. Specifically, they represent a major departure from the several equity principles, such as Egalitarian, Vertical, and Rawlsian, that were favored by the developing countries for many years.

The analysis is intended to offer guidance for policymakers negotiating and enforcing agreements to reduce GHGs in terms of promoting more equitable mitigation cost burden-sharing. It also identifies weaknesses in the Paris Accord that may cause it to unravel due to inequities in the pledges. Individual countries made pledges quite independently, but comparisons of relative positions will inevitably increase, and many developing countries will come to realize how much their prior bargaining positions have regressed.

4.2. Pre-Trading Equity Analysis

4.2.1. *Equity Principles*

In earlier work, we applied well-known principles of the Egalitarian, Ability to Pay, Vertical, and Horizontal equity, and the Rawlsian Maximin principle, for the study of the allocation of permits in GHG emissions protocols (Rose, 1992; Rose et al., 1998; Rose and Wei, 2008). This set of principles was extensively examined in the IPCC process, especially in the *Second Assessment*

[2] Given the slow progress in passing federal climate mitigation legislation in the US, emissions trading via individual states may in fact be the process by which the COP21 pledges are actually implemented in the near term.

Report (IPCC, 1995), but has been less central in subsequent reports. Still, many individual country proposals at the negotiating table referred to these principles.[3]

Building on our earlier work, we focus our analysis in the following sections on two major equity principles. The first equity principle is Egalitarian equity, which advocates for equal right of pollution or protection from pollution for each person. Based on this equity principle, the emission allowances are to be distributed on a per capita basis. The second equity principle is Ability to Pay, which relates the distribution of the mitigation burden (either measured by the total amount of emission reductions or the cost of the reduction effort) to the well-being or development level of the country/region. Based on this equity principle, the distribution of emissions reductions (usually measured as a percentage of baseline emissions) should be positively proportional to per capita GDP, and the cost of GHG mitigation (usually measured as a percentage change in GDP) should ideally be directly proportional to per capita GDP.[4]

4.2.2. *Basic Data*

At COP21, 195 counties and one regional government agreed to Intended Nationally Determined Contributions (INDCs) to reduce GHG emissions (UNFCCC, 2015). Table A.4.1 in the Appendix presents the GHG emission reduction pledges for the year 2030 for nearly one hundred countries that made *unconditional* pledges at COP21 that could be quantified in terms of percentage emission reductions (UNFCCC, 2016). The countries are listed in descending order of their absolute reduction pledge level. Also, some countries, mainly developing ones, made *conditional* pledges contingent on financial assistance from industrialized countries, but we confine our attention to only the unconditional pledges. Several countries included prospective emission allowance purchases toward their pledges, so we excluded these countries. Also, because some contributions could not be translated into precise pledges, we could not include all countries in the analysis. However, our sample includes countries with 88.1 percent of total projected GHG emissions in 2030, so we consider it

[3] More recently, a broader set of burden-sharing rules have been presented, not necessarily new in terms of equity principles as much as in terms of reference bases (Caney, 2012; IPCC, 1995; Averchenkova et al., 2014). They include such aspects as co-benefits of GHG mitigation and sustainability features. One such extension that has a more obvious equity implication is affordable energy, which has become a concern associated with the future path of allowance trading prices or a carbon tax (Victor et al., 2005).

[4] A distinction is sometimes made between the Ability to Pay and the Vertical equity principles, with the former only including the cost side and the latter including both the costs and benefits of mitigation. A related distinction is Ability to Pay being applicable to initial allocations of emissions allowances and Vertical equity being applicable to the outcome of the trading process as discussed later (see, Rose et al., 1998).

to be representative of the total set of parties to the Agreement (see Appendix A for the refinement of data used in our analysis).

4.2.3. *Index Analysis*

4.2.3.1. GINI COEFFICIENTS

In this section, we first use the Lorenz curve and Gini coefficient to analyze the equity implications of the COP21 pledges. Acknowledging the limitations of a one-parameter measure, the Gini coefficient characterizes the inequality of income distribution based on a comparison of the condition of a perfectly equal distribution of income among population and the actual distribution. There are alternative ways of calculating and interpreting the Gini coefficient when it is adapted to examining the equity implications of the distribution of GHG emissions or mitigation targets among countries (see Clarke-Sather et al., 2011; Teng et al., 2011, for an analysis of the inequality of GHG emissions; and Groot, 2010, for an analysis of hypothetical GHG mitigation agreements). In the analysis below, we calculate two versions of the Gini coefficient based on the data presented in Table A.4.1. One pertains to the distribution of GHG emissions with respect to population, and the other pertains to the distribution of pledged emissions reductions with respect to per capita GDP.

In Figure 4.1, the Lorenz curve is constructed by depicting the cumulative shares of population along the horizontal axis and the cumulative shares of

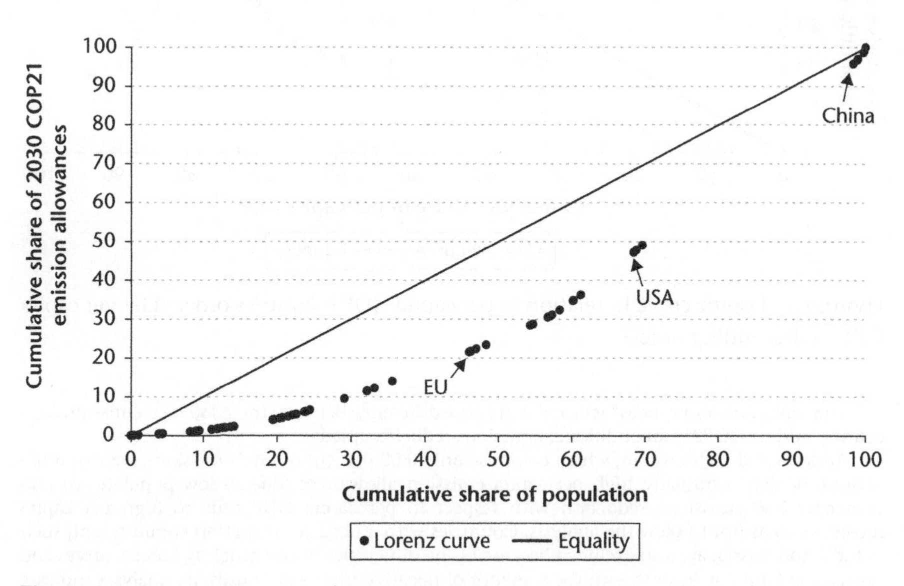

Figure 4.1. Lorenz curve in relation to population (countries ordered by per capita COP21 emission allowances in ascending order)

the 2030 "emission allowances" for each country along the vertical axis.[5] The countries are ranked in ascending order based on their respective per capita COP21 emission allowances.[6] Since the 45 degree Equality line represents the condition in which the per capita emission allowances are equal across the countries, this Lorenz curve reflects the departure of the emission allowances distribution from the perfect Egalitarian equity. Countries/regions that appear on the bottom end of the Lorenz curve, such as the EU, are in effect receiving a smaller share of emission allowances in relation to their population, while countries on the top end of the Lorenz curve, such as U.S. and China, are receiving a larger share of emission allowances relative to their population. The Gini coefficient calculated based on the Lorenz curve in Figure 4.1 is 0.323, which represents a considerable departure from the perfect Egalitarian equity.

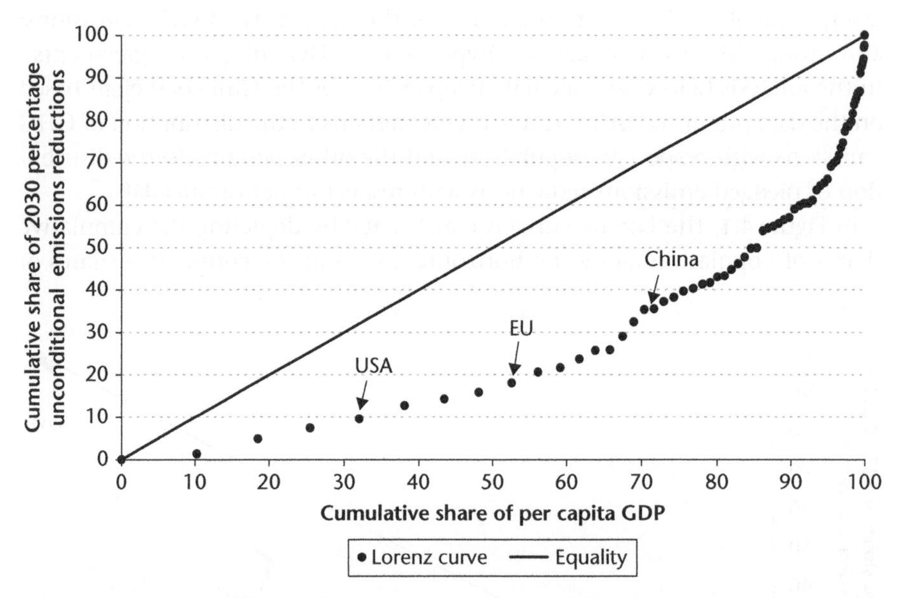

Figure 4.2. Lorenz curve in relation to per capita GDP (countries ordered by per capita GDP in descending order)

[5] The "emission allowances" is calculated as the difference between the 2030 BAU emissions of a country and its COP21 unconditional emissions reduction pledge.

[6] Andorra and Lichtenstein, which comprise only 0.02 percent of total emissions, were omitted because of their unusually high per capita emission allowances (due to low populations) and extremely low emissions reduction with respect to per capita GDP (due to high per capita incomes), so as not to skew the analysis. Countries with a negative reduction commitment, such as India and Russia, are also excluded because of the difficulties in constructing Lorenz curves and calculating Gini coefficients with the presence of negative numbers. Sensitivity analyses indicate that the exclusion of these countries (which reduces world emissions coverage from 88 percent to 75 percent) only results in very small changes to the Gini coefficients (within a few percent).

The Lorenz curve in Figure 4.2 is constructed based on the Ability to Pay equity. The vertical axis depicts the cumulative share of COP21 unconditional pledged GHG reductions in percentage terms.[7] The horizontal axis depicts the cumulative share of per capita GDP. The countries are ranked in descending order based on their respective per capita GDP. The 45 degree perfect equality line in this case represents a direct proportional relationship between pledged percentage emissions reductions and per capita GDP, implying that wealthier countries should proportionally take more emissions reductions in percentage terms. Countries that appear on the bottom end of the Lorenz curve, such as the US, are making relatively low emissions reduction pledges in relation to their per capita GDP; while countries on the top end of the Lorenz curve, largely lower-income countries, are making more emissions reduction pledges relative to their per capita GDP. The Gini coefficient calculated based on the Lorenz curve in Figure 4.2 is 0.515, which represents a high level of inequality and a significant departure from Ability to Pay equity.

4.2.3.2. ATKINSON INDEXES

We also computed the Atkinson index for the sample of countries in Table A.4.1. This index is considered superior to the Gini coefficient though it is less immediately transparent. It is also a single-parameter measure of inequality, but it has a stronger basis in economic theory, particularly welfare analysis (Atkinson, 1970). The key parameter here is the "degree of inequality aversion," which represents the weight society places on distributional inequality. As the value of the parameter increases, the Atkinson index becomes more sensitive to changes in the income of the population at the lower end of the distribution (Braun, 1988; Du et al., 2015). The original Atkinson measure is based on a utility function of the following form:

$$U(y) = \begin{cases} \dfrac{y^{1-\epsilon}}{1-\epsilon} & \epsilon > 0, \epsilon \neq 1 \\ ln(y) & \epsilon = 1 \end{cases} \tag{1}$$

where y is personal income and ϵ is the inequality aversion parameter. When y increases, utility increases at a diminishing rate.

To analyze Egalitarian equity using this metric, we replace y with E, which represents the per capita emission allowances as a result of COP21.[8] We believe it is reasonable to assume that the utility function of per capita emission allowances has similar properties as the utility function in equation

[7] This is derived by calculating the emissions reduction percentage in each country, then summing them, and finally rescaling to 100.

[8] This is again computed for each country by subtracting the unconditional committed emissions reduction target from the BAU emission level in 2030.

(1), i.e. utility increases as people are granted more emission allowances but at a decreasing rate (diminishing marginal utility). The Atkinson index (*AI*) is now defined by equation (2):

$$AI = \begin{cases} 1 - \left[\sum_{i=1}^{n} \left(\frac{E_i}{\bar{E}_i}\right)^{1-\epsilon} P_i\right]^{\frac{1}{1-\epsilon}} & \epsilon > 0, \epsilon \neq 1 \\ 1 - \prod_{i=1}^{n} \left(\frac{E_i}{\bar{E}_i}\right)^{P_i} & \epsilon = 1 \end{cases} \quad (2)$$

where E_i is the per capita emission allowances for country *i*, $\bar{E}$ is the average per capita emission allowances of all the sample countries, P_i is the population share of country *i*, and *n* is the total number of countries in the sample. The value of the *AI* varies between zero and one, where the former indicates zero and the latter indicates infinite social utility can be gained by a redistribution of the entitlement (in this case, emission allowances) so that each individual has the same level of entitlement. Therefore, a lower value of *AI* represents a more equal distribution than a higher value of *AI*. The Atkinson index can also be interpreted in this way: for an *AI* that equals 0.3 for a given ϵ, if every country is granted the same level of per capita emission allowances, only $1 - AI = 0.7$ (or 70 percent) of the total emission allowances is needed to achieve the same level of social welfare that is obtained by the current distribution of the emission allowances (Hedenus and Azar, 2005).

In Table 4.1, we compute the *AI* for both the per capita emissions in the Business-as-Usual (BAU) scenario and the per capita emissions as a result of the COP21 pledges. The results indicate that the COP21 pledges result in higher Atkinson indexes or higher distributional inequality at all the values of the inequality aversion parameter ϵ we examine. Essentially, the results indicate that COP21 is not consistent with Egalitarian equity in terms of the distribution of GHG emissions allowances.

4.2.4. *Equity Principle Analysis*

We can also gain insight into the COP21 INDCs by comparing them more directly with established equity principles. There is a general consensus among researchers about the equity principles relevant to international climate

Table 4.1. Comparison of Atkinson index between the BAU scenario and COP21 scenario

ϵ	0.2	0.5	0.8	1	1.2	1.5	1.8	2.0
BAU scenario	0.0347	0.0915	0.1548	0.2008	0.2494	0.3258	0.4029	0.4523
COP21 scenario	0.0395	0.1038	0.1746	0.2251	0.2776	0.3581	0.4370	0.4866

negotiations (see, e.g., IPCC, 1996; Kverndokk and Rose, 2008; Averchenkova et al., 2014; Fair Shares, 2015). Supplemental insight can be obtained by examining correlations between key variables as presented in Table 4.2. If we examine the correlations in terms of the pledged percentage reductions in GHG emissions, the correlation between it and per capita GDP is relatively low. The correlation between percentage GHG mitigation pledges and population is negative, though near zero. Pairwise regression analyses further suggest that the correlation between population and unconditional GHG reduction target is not statistically different from zero.

Three of the most popular equity principles are Horizontal, Ability to Pay, and Egalitarian equity. The first calls for all countries to have the same percentage emissions reduction requirements. At this point, we can clearly see that the Horizontal equity principle is not operative in terms of the allocation of emissions reduction pledges. The Ability to Pay principle would call for countries with higher per capita GDP to commit to higher percentage GHG reductions. However, the very low correlations indicate it is not operative here either.[9]

As to the Egalitarian principle, historically this was favored strongly by developing countries, which usually have large populations and low per capita GDPs, and hence per capita emissions. In previous negotiations it meant that more populous countries would receive the lion's share of GHG emission allowances. Essentially, this would be consistent with an inverse correlation between population and GHG emissions reduction requirements, but the results in Table 4.2 indicate only a very weak inverse correlation. Thus, we can conclude that COP21 resulted in a *shift* in the position of many developing countries. Essentially the position of populous developing countries in

Table 4.2. Correlation of key variables

	2030 unconditional GHG reduction target (in percentage terms)	Population	BAU emission	GDP	Per capita GDP
2030 unconditional GHG reduction target (in percentage terms)	1				
Population	−0.042	1			
BAU emission	−0.042	0.970	1		
GDP	0.088	0.753	0.764	1	
Per capita GDP	0.366	0.010	0.049	0.172	1

[9] The correlation between countries' absolute levels of GDP and emissions reduction level targets is very high (at 0.867), but this is not really related to either of the two aforementioned equity principles, but is rather more a function of country size.

prior years was that they would receive allowances in excess of their emissions, which implicitly meant a negative emissions reduction requirement. However, these same countries might have found it embarrassing to state their pledges in these terms at COP21. Moreover, the traditional position of industrialized countries to fend off being subject to the Vertical or Ability to Pay equity principles seems to have been attained in Paris when one looks at the low correlation between per capita GDP and emissions reduction pledges. Although pledges have been made, history has shown that not all countries view them as binding. In recent years, Canada and Australia, for example, have backed off from their Kyoto commitments. Knowledge of the equity implications of the COP21 pledges will be helpful in any future multilateral revisions of the pledges, or in dealing with countries who threaten to back out of them.

4.3. Emissions Trading Model

A GHG emissions permit (allowance) trading system between the three countries/regions is simulated using a variant of the nonlinear programming modeling framework of Rose et al. (1998) and Rose and Wei (2008). The model is of the following form, beginning with the objective function, which calls for the minimization of GHG mitigation costs for the three countries/regions as a whole:

$$\text{minimize} \quad TC = \sum_{i=1}^{n} \left[\frac{a_i}{b_i} e^{b_i R_i} - \frac{a_i}{b_i} - a_i R_i \right] \cdot E_i \qquad (3)$$

where

$TC \equiv$ total mitigation cost of the three countries/regions (endogenous)

$R_i \equiv$ percentage mitigation for country i (endogenous) $0 \leq R_i \leq 1$

$E_i \equiv$ gross (unmitigated) GHG emission (in tons of CO_2e) for country i (exogenous)

$n = 3$

a_i and $b_i \equiv$ intercept and slope parameters of mitigation cost function for country i (exogenous)

mitigation marginal cost function of the form: $a_i e^{b_i R_i} - a_i$

subject to the following constraints, where the emissions of any country/region must be less than or equal to the sum of its initial emission allowance allocation and its allowance purchases or sales:

$$(1 - R_i)E_i \leq \bar{P}_i + P_i \quad i = 1 \ldots n \qquad (4a)$$

and where the sum of all allowance purchases and sum of all allowance sales in the market must be equal (i.e. their sum must be equal to zero):

$$\sum_{i=1}^{n} P_i = 0 \tag{4b}$$

where
$\bar{P}_i \equiv$ allowance allocation for country i (exogenous)
$P_i \equiv$ allowance purchases or sales (in tons) for country i (endogenous)

4.4. Emissions Trading Analysis

The analysis thus far has been limited to initial GHG reduction pledges, but this leaves open the issue of how they will actually be implemented. An emissions trading system is considered to be a likely feature of the realization of the pledges. The cap has essentially been established by the pledges, and trading traditionally has been viewed as a way that the global commitment can be achieved at least cost. But accordingly, it also means that each country can lower its cost by either buying or selling emission allowances, or permits. We now turn to whether these transactions will moderate or exacerbate the inequities of the pledges identified in Section 4.3.

We perform our emissions trading analysis with just a handful of representative countries/regions for two reasons. First, mitigation cost curves are not available for all countries that made unconditional pledges at COP21. Second, an analysis for all of countries would lead to some unwieldy tables of results. Hence, we confine our attention to some representative countries/regions. The EU is a representative collection of high-income countries and is itself as a signatory to the Paris Agreement. China is a major emitter of GHGs, and, although it has made great progress in recent years, is still a developing country. California is a region within a larger country, and also representative of how progress might be made toward achieving COP21 goals—it has committed itself to these goals despite the US withdrawing from the Paris Agreement. We also bring India into the analysis as a representative developing country in terms of its mitigation cost curve.

4.4.1. *Basic Data*

The model is applied to the basic data in Table 4.3 and to GHG mitigation cost curves for the trading entities, the basic form of which are presented in Figure 4.3. Our analysis will modify these curves of macroeconomic considerations and alternative future scenarios for technologies having a major influence on GHG emissions.

Two widely used modeling approaches to develop marginal abatement (mitigation) cost (MAC) curves are the bottom-up approach and top-down approach. The former relies on the methodological framework of "engineering-cost,"

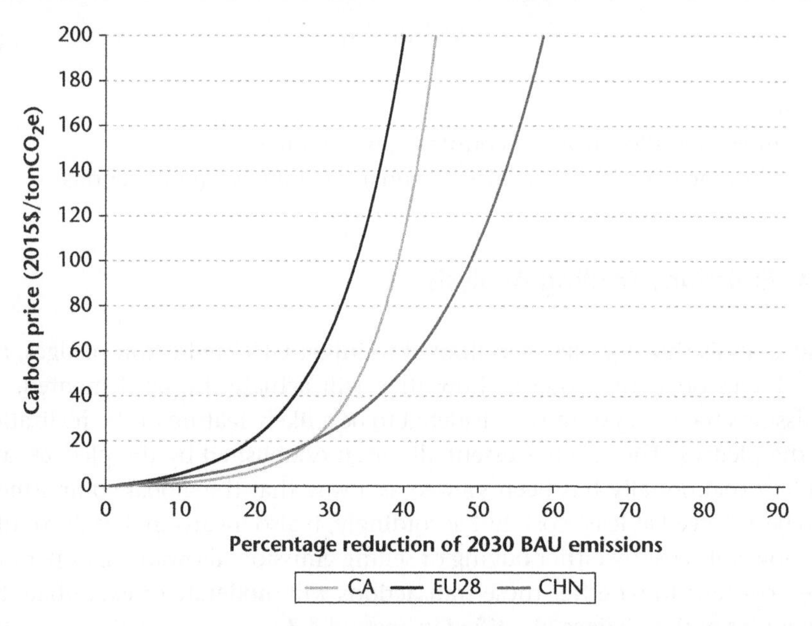

Figure 4.3. Marginal cost curves of GHG mitigation for California, EU, and China in 2030

Table 4.3. Basic data

Trading parties	BAU gross emissions in 2030 (million tonCO_2e)	Emissions cap in 2030 (million tonCO_2e)	GHG mitigation goal in 2030 (relative to BAU emissions) (%)	Autarkic marginal mitigation cost ($ per tonCO_2e)
California	542	260	52.1	627.5
EU	4,661	3,141	32.6	94.8
China	19,721	19,253	2.4	0.6
Total	24,924	22,649	9.1	

which incorporates substantial technological details of individual mitigation policy options. They are typically partial equilibrium analyses that operate through the objective of cost minimization. The top-down approaches primarily utilizes macroeconomic simulation models that include substitution effects, economy-wide interactions, and other macroeconomic feedbacks from changes in price, income, and trade. Examples of bottom-up models include MACs for fifteen countries developed by McKinsey & Company (2013). Top-down models include the Emission Prediction and Policy Analysis (EPPA) model by Ellerman and Decaux (1998), later updated in Morris et al. (2012), and the Dynamic Integrated Climate-Economy (DICE) model by Nordhaus (Nordhaus and Sztorc, 2013). The top-down models are often accused of lacking technological details and overestimation of marginal mitigation costs, while the bottom-up models are often accused of underestimation of the costs due primarily to the overestimation of cost-saving mitigation opportunities (Rose et al., 2009; Kesicki, 2011; Kesicki and Ekins, 2012).

In this study, the *microeconomic* marginal abatement cost curves for the EU and China are derived from simulations with the GEM-E3 computable general equilibrium model, a top-down approach (see Capros et al., 2013, for an extensive description of the model and Vandyck et al., 2016, for a recent application to an assessment of international climate policy). The simulations impose a range of global carbon prices uniformly across a given region up to the year 2030. The parameters a_i and b_i of an exponential functional form for the curve are estimated using the ordinary least squares regression technique to fit the simulation results for each region *i*.[10] The MAC has the following functional form:

$$MAC_i = a_i e^{b_i R_i} - a_i$$

where R_i is the level of abatement relative to a reference scenario with zero carbon price up to 2030.

The California curve was extrapolated from a study of southern California (SCAG, 2012; Wei and Rose, 2014), which was based on a careful analysis of mitigation potential and costs by government agencies and an interdisciplinary team of researchers and stakeholders. In order to use the cost data in our NLP model, we use regression analysis to fit a smooth curve through a step function that represents the mitigation potential and cost associated with individual policy options. The fitted curve has the same functional form as the EU and China cost curves derived from the GEM-E3 Model. To be consistent with these other MACs, we force the fitted California curve through the origin as well.

4.4.2. *Base Case Analysis*

The results of our base emissions allowance trading case are presented in Table 4.4. Before trading, China incurs very minimal costs because of its very low emissions reduction target and relatively flat cost curve. California, despite having a flatter cost curve than the EU, incurs the highest pre-trading mitigation cost, primarily due to California's high emissions reduction target.

[10] Note that McKinsey & Company (2013) and many other bottom-up studies indicate substantial negative cost mitigation potentials. Our compromise is to drop the positive cost intercept for the GEM-E3 MACs, thereby running the curves through the origin.

Table 4.4. Economy-wide emissions trading simulation among California, the EU, and China in year 2030

Trading parties	Before trading	After trading			Cost saving	Allowances traded	Emissions reduction after trading		Original emissions reduction goal
	Mitigation cost	Mitigation cost	Trading cost[a]	Net cost		(million tonCO_2e)	(million tonCO_2e)	(% from BAU)	(% from BAU)
California	24,679	90	666	756	23,923	202	80	14.8	52.1
EU	37,131	459	3,964	4,423	32,708	1,203	316	6.8	32.6
China	146	2,753	−4,629	−1,876	2,022	−1,405	1,879	9.5	2.4
Total	61,957	3,303	0	3,303	58,654	1,405[b]	2,275	9.1	9.1

[a] Allowance price = $3.29/ton$CO_2$e.
[b] Represents the number of allowances bought or sold.
Note: Based on microeconomic cost curves (million dollars or otherwise specified).

The total mitigation cost for the three entities before trading is $62 billion. Trading results in about 95 percent cost saving, and lowers the total cost of achieving the same aggregate emissions reduction goal to just $3.3 billion in 2030, which is consistent with the implications of the Coase theorem. China sells 1,405 million allowances, most of which are purchased by the EU at an equilibrium price of $3.29/ton$CO_2$e. Accordingly, the EU obtains greater cost savings ($32.7 billion) from its allowance purchases than does California ($23.9 billion). Even though the allowance purchase by California enables it to reduce its autarkic emissions reduction from 52.1 to 14.8 percent, while the EU reduces its autarkic emissions reduction from 32.6 to 6.8 percent, the latter reaps greater cost savings because its marginal mitigation cost curve is relatively steeper. In fact, the EU's net gains are 37 percent greater than that of California. Moreover, the EU's gains are more than sixteen times those of China, primarily because the margin between the allowance price and its increased mitigation costs (due to its need to back up allowance sales) is rather low for the latter entity.

4.4.3. *Macroeconomic Cost Analysis*

Many analyses are performed with engineering-based, or microeconomic, GHG mitigation cost curves (see, e.g., Rose and Stevens, 1993; Ellerman and Decaux, 1998). However, given strong interdependencies in most nations' economies and the significant cost of mitigation, this approach is likely to understate the total impacts of mitigation. For example, studies have found that urban forestry (tree planting to sequester carbon) has a relatively high direct cost per unit mitigated, while a renewable portfolio standard (RPS) has a relatively lower, but still positive, cost. Because an RPS pertains to electricity production, the increased cost ripples through the economy and multiplies several-fold, leading to a potentially significant negative macro impact (Wei and Rose, 2014). On the other hand urban forestry is at the very end of the production stream and does not conjure a demand price, and hence has very limited multiplier or general equilibrium effects. Wei and Rose (2014) indicate that the relative positioning of these two options switches when one considers the macro implications.

In this section, we perform a sensitivity analysis on emissions allowance trading using macroeconomic cost curves for the EU, China, and California. The adjusted macroeconomic cost curves for the three trading entities are presented in Figure 4.4. For the first two, we use the macro-level MACs from the GEM-E3 model. A similar adjustment was made to the MACs as for the micro-level cost curves, i.e. to adjust the cost curves downwards to pass through the origin. The macroeconomic cost curve for California was developed based on the relationship between the macro- and micro-level cost curves developed in Wei and Rose (2014) for the southern California region. Marginal cost curves at both the micro and macro levels are constructed and compared for the study region, and the macro to micro ratios for both the intercept coefficient and the slope coefficient in the marginal cost functions are computed. These ratios are applied to the coefficients in the micro marginal cost functions for California to obtain the corresponding macro cost curves as depicted in Figure 4.4.

Table 4.5 presents the results of emissions trading based on the macro-level cost curves. When we take into consideration the total cost (including both direct and indirect costs) of mitigation options, the equilibrium allowance price increases from \$3.29/tonCO_2e in the Base Case to \$14.41/tonCO_2e in this case. China again is the allowance seller, while the EU and California are the buyers. California obtains greater cost savings from its allowance purchases than does the EU in both absolute and relative terms. Both of them obtain cost savings from trading greater than China in absolute terms. However, China's net cost savings also include \$9.7 billion in revenues from allowance sales.

Compared to the microeconomic cost results presented in Table 4.4, the results of taking macroeconomic considerations into account indicate greater cost savings from trading, more allowances traded, and greater per unit gains from trading. Of course the overall emissions reductions are the same, but not surprisingly the emissions reductions by California and the EU have decreased

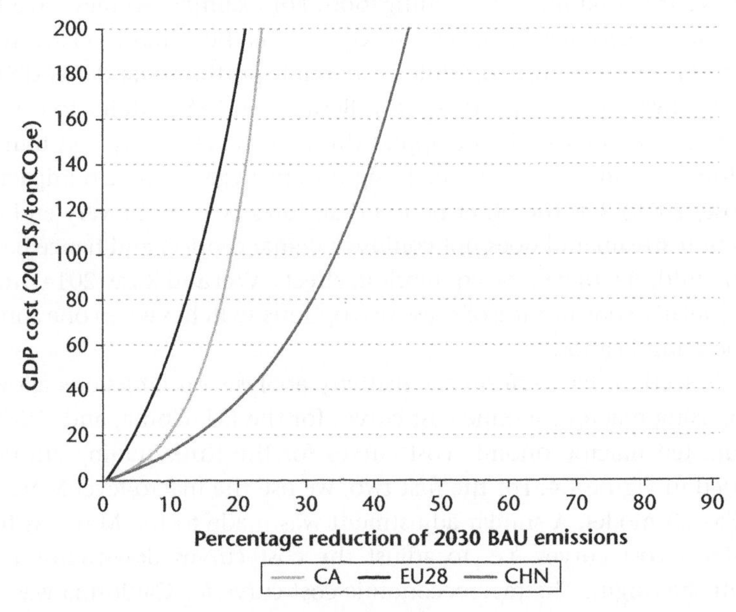

Figure 4.4. Macroeconomic marginal cost curves for China, EU, and California in 2030

Table 4.5. Economy-wide emissions trading simulation among California, the EU, and China in year 2030

Trading parties	Before trading	After trading			Cost saving	Allowances traded	Emissions reduction after trading		Original emissions reduction goal
	Mitigation cost	Mitigation cost	Trading cost[a]	Net cost		(million tonCO_2e)	(million tonCO_2e)	(% from BAU)	(% from BAU)
California	550,314	243	3,471	3,715	546,599	241	41	7.6	52.1
EU	267,394	980	19,856	20,836	246,558	1,378	141	3.0	32.6
China	591	13,590	−23,327	−9,737	10,328	−1,619	2,093	10.6	2.4
Total	818,299	14,813	0	14,813	803,486	1,619[b]	2,275	9.1	9.1

[a] Allowance price = $14.41/ton$CO_2$e.
[b] Represents the number of allowances bought or sold.
Note: Based on macroeconomic cost curves (million dollars or otherwise specified).

and those of China have increased relative to the results of the microeconomic cost curve case. This is primarily because the cost curves for the EU and US become even steeper relative to China's curve than in the micro-level case. For completeness, we mention the Green Climate Fund transfers, primarily from industrialized countries who committed to the COP21 Accord, totaling approximately $100 billion, to go to developing countries. However, these pertain to the conditional pledges made in Paris, and we have confined our focus to the unconditional pledges, for which the transfers are less relevant.

4.4.4. *Analysis of Alternative Technology Paths*

Advances in the extraction, transformation, and utilization of primary energy resources and of GHG mitigation and sequestration options have potentially prominent implications for the direct cost of achieving climate policy goals. The future of technological changes is highly uncertain, but sensitivity analyses can be performed to analyze the potential aggregate and distributional impacts across countries/regions. We examine the equity implications of one major technology alternative.

Figure 4.5 presents the cost curves for incorporating the full potential of fuel economy standards into California's mitigation cost curve. We base

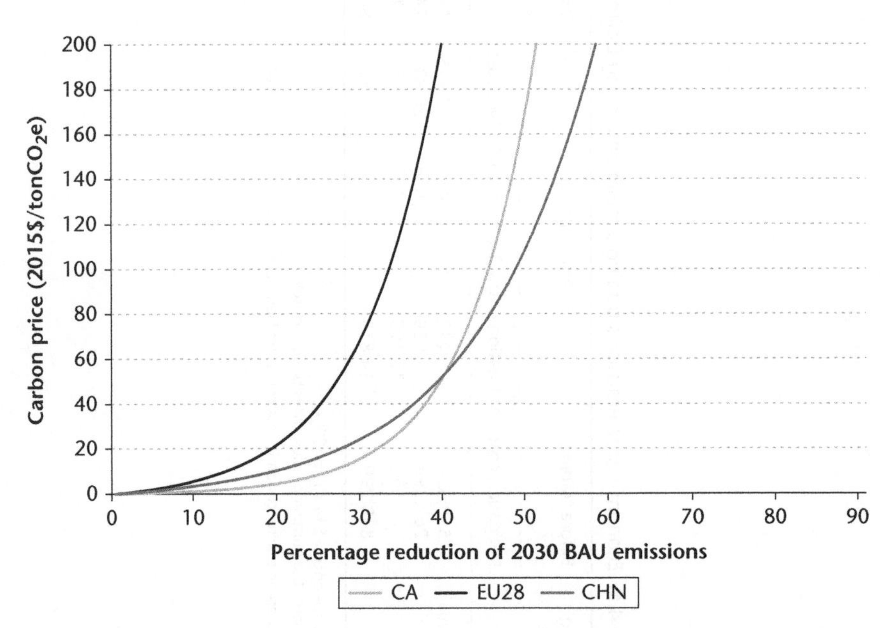

Figure 4.5. Marginal cost curves of GHG mitigation for California, EU, and China in 2030 (including fuel economy standards in California's MAC)

our cost and mitigation potential estimates on the following assumptions and analysis:

- In 2004, the California Air Resources Board (CARB) approved fuel economy standards to reduce emissions from new passenger vehicles and light trucks, beginning with 2009 model year (the so-called "Pavley" standards).
- Assuming no leakage, the emissions reductions potential of the standard are 30 million metric tons by 2020 and over 50 million metric tons in 2030. To put this in perspective, this represents an overall reduction of 18 percent of the passenger car emissions by 2020.
- Importantly, these standards also generate auxiliary benefits in the form of reductions of ozone-forming emissions, including hydrocarbons and nitrogen oxides. The California Air Resources Board estimates reductions of 6 tons per day in 2020.
- In calculating the cost per ton of emissions reductions, one needs to consider whether consumers value fuel savings or not, as this substantially changes the costs.
  - If consumers value fuel savings for three years, the resulting cost is \$30/tonCO_2
  - Typically, CARB assumes that consumers value fuel savings over fourteen years; in this case the cost is actually negative, at −\$135/tonCO_2
- We compare these estimates against the academic literature on fuel economy standards. Anderson and Sallee (2011), using "loopholes" to reveal the marginal cost of regulation, estimate that tightening standards by one mile per gallon would cost \$9–27. This is more aligned with the idea that consumers value fuel economy savings for three to five years, but not over fourteen years.
- In our analysis, we use the average of the upper- and lower-bound per ton cost estimates from Sweeney (2016).

Table 4.6 presents the results of emissions trading based on the technology path adjusted cost curves. Compared to the Base Case, the results change only marginally in terms of the equilibrium allowance price and allowances traded. This is primarily because California has much fewer BAU emissions compared to China and the EU, and thus a change in California's MAC only has a relatively small influence on these results. The allowance price decreases only from \$3.29/tonCO_2e to \$3.27/tonCO_2e. California increases its own mitigation from 14.8 percent to 17.4 percent, and thus reduces allowance purchases by about 14 million tons. The total allowances traded decrease from 1,405 million tons to 1,393 million tons. Two major changes from the Base Case take place. First, California's cost savings decrease from \$23.9 billion to \$10 billion, primarily because, with a flatter cost curve, the state has a much lower mitigation

Table 4.6. Economy-wide emissions trading simulation among California, the EU, and China in year 2030

Trading parties	Before trading	After trading			Cost saving	Allowances traded	Emissions reduction after trading		Original emissions reduction goal
	Mitigation cost	Mitigation cost	Trading cost[a]	Net cost		(million tonCO_2e)	(million tonCO_2e)	(% from BAU)	(% from BAU)
California	10,673	106	614	720	9,954	188	94	17.4	52.1
EU	37,131	453	3,935	4,388	32,743	1,205	314	6.7	32.6
China	146	2,714	–4,549	–1,836	1,982	–1,393	1,867	9.5	2.4
Total	47,951	3,272	0	3,272	44,679	1,393[b]	2,275	9.1	9.1

[a] Allowance price = \$3.27/tonCO_2e.
[b] Represents the number of allowances bought or sold.
Note: Based on technology-adjusted microeconomic cost curves (million dollars or otherwise specified).

cost before trading ($10.7 billion vs. $24.7 billion in the Base Case). Second, the total cost savings of the three entities reduces from $58.7 billion to $44.7 billion, with the majority of the change associated with California (ironically, the state's net cost after trading, which is the sum of mitigation cost and trading cost, hardly changes at all). The EU ends up with slightly higher cost savings and China with slightly lower cost savings from trading.

4.4.5. *GHG Emission Offsets*

The use of GHG emissions offsets represents another alternative to reduce the costs of achieving COP21 targets and possibly to improve equity. An "offset" refers to an emission reduction from a "non-covered" source (Bento et al., 2015). This could be a sector within a country that made a COP21 pledge, or it could be a country that did not sign the Paris Agreement. Technically, there is one other possibility, which would be a country that signed the Agreement but whose year 2030 pledge is such that it is not required to undertake any mitigation. The advantages of including offsets in the policy design are to bring in lower-cost mitigation options. If these options are purchased from low-income countries, and a large number of such options do exist there, it holds the possibility of their obtaining additional financing through allowance sales.

In our analysis, we assume that India, since it has a negative emission reduction pledge for the year 2030 under COP21, would be able to sell emissions reductions as offsets in the emissions trading market. We use the microeconomic MAC for India derived from the GEM-E3 model in the simulation. Figure 4.6 presents the cost curves for the four trading entities in this case. India's MAC is actually higher than that of both California and China at first, but then crosses beneath California's curve at around the 35 percent level.

Table 4.7 presents the simulation results. Inclusion of India makes additional low-cost emissions reductions available. This lowers the allowance price from $3.29/ton$CO_2$e in the Base Case to $2.60/ton$CO_2$e in this case. Total cost savings slightly increase from $58.7 billion to $59.3 billion. With a lower allowance price, the EU and California mitigate fewer emissions on their own and purchase more from the market. This results in a slight increase in their respective cost savings. A total of 1,464 million tons of allowances are traded. China's cost savings decrease (from $2 billion to $1.2 billion), as 25 percent of the allowances are now sold by India.

4.5. Post-Trading Equity Analysis

We now compare the post-trading equity outcomes with the pre-trading equity outcomes relating to unconditional INDCs. Although we only simulate

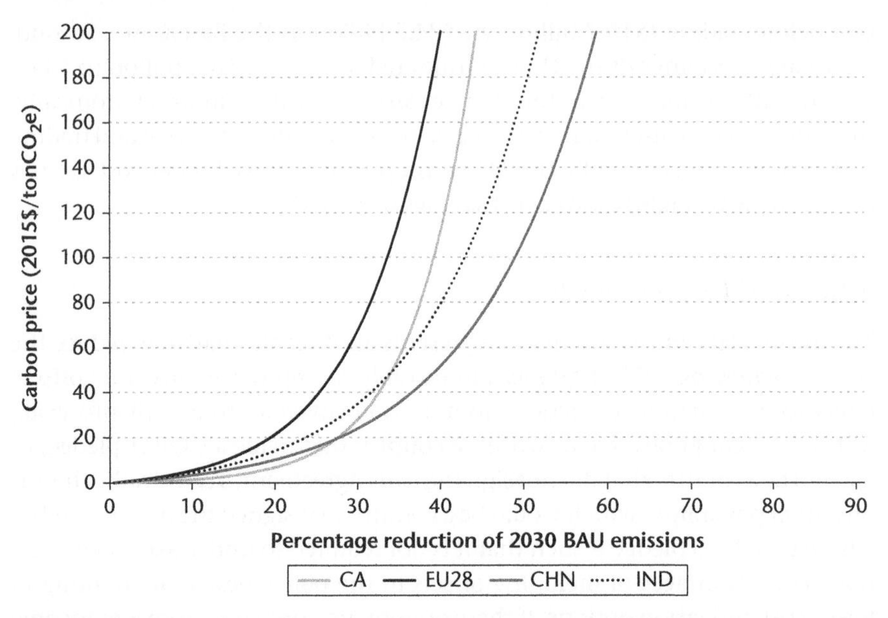

Figure 4.6. Marginal cost curves of GHG mitigation for California, EU, China, and India in 2030

three countries/regions, we can gain some insights that enable us to generalize our results to the seventy-five countries in the Appendix (Table A.4.1) that have made positive unconditional GHG reduction pledges.

Focusing first on the results in Table 4.4, our Base Case, the countries/regions with the highest current GDPs reap the greatest rewards of trading. California's cost saving is nearly $24 billion, while the EU's cost saving is over $32 billion. China only obtains a cost saving of $2 billion. For the macroeconomic case in Table 4.5, California's and the EU's cost savings are $547 billion and $247 billion, respectively, but China's total cost saving is only $10 billion. The predominant shares of the post-trading gains go to the more well-off regions again, counter to Ability to Pay equity. These results could generalize to all seventy-five countries because California and the EU are representative of the vast majority of the other industrialized countries in our sample in terms of level of economic development, gross emissions, mitigation cost curves, and absolute and percentage mitigation pledges (see Rose et al., 2017). China is atypical of developing countries because of its larger GDP, high GHG emissions, relatively flat MAC, and relatively low emissions reduction pledge. However, since China is predicted to capture a lesser proportion of the gains from emissions trading than California and the EU despite these relative advantages, most of the other developing countries are likely to do so as well, hence scoring low on both Ability to Pay and Egalitarian equity.

Table 4.7. Economy-wide emissions trading simulation among California, the EU, China, and India (offset) in year 2030

Trading parties	Before trading	After trading			Cost saving	Allowances traded	Emissions reduction after trading		Original emissions reduction goal
	Mitigation cost	Mitigation cost	Trading cost[a]	Net cost		(million $tonCO_2e$)	(million $tonCO_2e$)	(% from BAU)	(% from BAU)
California	24,679	67	546	613	24,066	210	72	13.3	52.1
EU	37,131	311	3,261	3,572	33,559	1,254	266	5.7	32.6
China	146	1,858	−2,864	−1,006	1,152	−1,101	1,574	8.0	2.4
India	0	434	−943	−509	509	−363	363	6.4	0.0
Total	61,957	2,670	0	2,670	59,287	1,464[b]	2,275	7.4	7.4

[a] Allowance price = $2.60/$tonCO_2e$.
[b] Represents the number of allowances bought or sold.

Note: Based on microeconomic cost curves (million dollars or otherwise specified).

On a per capita GDP basis, California's gains are the highest in the macro cost curve case and the EU's gains are the highest in the other three cases, while China's gains are the lowest in all four cases. Again, this means a worsening of Ability to Pay equity. The results also have negative implications for the Egalitarian equity principle. The gains are inversely correlated with population and are positively correlated with per capita GHG emissions. Again, these results can be generalized to all of the seventy-five countries in our sample.

For the alternative technology scenario presented in Table 4.6, the equity results are hardly affected despite the fact that changes in the availability of fuel economy technology would have a differential effect across the three regions. This is likely to be the case for any other single technological change, since few individual technologies outside of electricity generation and motor vehicles generate more than 10 percent of GHGs.

For the case of including offsets shown in Table 4.7, both Egalitarian and Ability to Pay equity are worsened in relation to the Base Case. The presence of offsets lowers the price of allowances by 21 percent, and, with only a very slight increase in allowances traded, means that the combined revenue gains by China and India are nearly $1 billion lower. Moreover, the overall cost savings for California and the EU are both slightly higher, and the overall cost savings for China and India are 18 percent lower. As India is fairly representative of other developing countries, a similar qualitative result is likely to hold for the inclusion of more of them into the trading system via offsets.

4.6. Further Analysis

Three additional perspectives are presented on the equity of cross-country INDCs and their stability. First is the prospect that developing countries were more likely to offer higher pledges at COP21 because of the promised large financial transfers from industrialized countries, specified as $100 billion annually. This is one of the reasons we have focused on *unconditional* pledges. However, even these are a major departure from previous commitments by developing countries. Conditional pledges for the Paris Accord are even more ambitious and thus a greater departure.[11] Some analysts have suggested that much more than $100 billion is needed annually (World Bank, 2010). Moreover, actual commitments of these transfers to date are only a fraction of the total (Green Climate Fund, 2016).

[11] Given some ambiguity in the language of the Paris Agreement, it is also possible that some developing countries would receive transfers even for their unconditional commitments.

Will emissions trading provide the necessary capital to make it feasible or political viable for developing countries to honor their Paris pledges, not to mention their conditional ones? The analysis in Section 4.5 indicates that China would receive only between $1.2 billion and $10.3 billion in emission allowance revenues annually. Expansion of the analysis to include more countries is needed to make a complete assessment, however. In addition, this analysis only pertains to the initial year of the pledges, and the percentage emission reduction targets are likely to continue to move the Chinese economy up its steepening marginal cost curve as economic growth continues.

Another perspective on the equity of cross-country INDCs and their stability is offered by Averchenkova and Bassi (2016), who examine the credibility of the INDCs for the G20 countries, defining credibility as the "likelihood that policymakers will keep promises to implement their pledges" (p. 3). They note that "most countries also indicated in their INDCs why they consider their intended contribution to be fair and ambitious in the global context" (p. 6). It is not clear, however, to what extent individual countries evaluated their pledges relative to others, since commitments and pledges were a moving target leading up to COP21. In any case, it would not be unusual for there to be something akin to buyer's remorse in the aftermath of the Accord. Adding to the concern about the tenuous nature of the Accord, Averchenkova and Bassi have assessed the pledges by the sample of G20 countries as only moderately credible on average.

4.7. Conclusion

We have analyzed the equity implications of the COP21 Agreement negotiated in Paris in December 2015. Our analysis was performed in two parts. First, we analyzed the initial *allocation* of voluntary GHG reduction pledges directly (before any emissions allowance trading) by using correlation analysis, and computing Gini coefficients and Atkinson indexes. Then we simulated the effect of trading for a small sample of representative countries/regions to analyze the *outcomes* of the Paris Agreement in terms of cost implications in relation to country GDP and population.

Our analysis indicates that the GHG reduction pledges made at COP21 run counter to major equity principles such as Ability to Pay, Vertical, Horizontal, and Egalitarian equity. They are definitely a major departure from the Egalitarian, Vertical, and Rawlsian equity principles proposed for many years by many developing countries. The results give rise to concern that many of the signatories of COP21 might not follow through on their unconditional pledges. Ironically, while emissions trading reduces the global cost of mitigating GHGs and provides much-needed transfers to low-income countries through allowance sales revenues, wealthier countries stand to gain a much higher proportion of the savings, thereby worsening the equity outcome.

Appendix A. Basic Data

Table A.4.1 presents the GHG emission reduction pledges for year 2030 for nearly one hundred countries that made *unconditional* pledges at COP21 that could be quantified in terms of percentage emission reductions. Pledges for several counties needed to be translated from other base years (e.g. 15 percent below year 2005 levels) and adjusted to year 2030 emission projections.

When calculating each country's pledge to reduce emissions, data were first collected from their INDC as submitted for the COP21 proceedings. This included their total GHG emissions in the base year, projected BAU emissions, and any amount of pledged reduction. The pledges were then standardized as both a total and percentage reduction from a 2030 BAU scenario. Conditional and unconditional pledges were calculated separately as indicated in the country's INDC. In Table A.4.1, we only present the unconditional pledge data that are used in this analysis.

Given that not all countries conformed to the same baseline or projected year for their BAU emissions, nor the same way of measuring emissions or reductions, we first standardized each county to a 2030 BAU projection from which we then calculated their emission reduction target in both absolute and percentage terms. For example, if a country measured their emissions intensity ($mtCO_2e$/GDP) or emissions per capita, we used current and 2030 projected GDP and population numbers to translate these into absolute and percentage terms. In some cases, the GDP and population projections were included in the INDC; in cases where they were not, projections were adopted from the US Census Bureau (2015) and Live Population (2016) for population, and the US Department of Agriculture's Economic Research Service (2015) for GDP.

Additionally, if a country provided its information in a year pre-2030, we assume that the pledge would hold as is through 2030. If, however, a country provided their information in a post-2030 year, we linearly extrapolated backward using either their current baseline emissions, or, if provided, an intermediate goal set forth in their INDC. For example, some countries offered both a final 2035 goal and an intermediate 2025 goal, which were averaged to calculate their approximate 2030 goal.

Further complicating the issue, a few countries' INDCs did not provide exact numerical data, only graphs of their emissions and reduction levels. In these cases we used our best judgment to infer values from the graphs when they were drawn to scale. If no numbers or graphs were reported for baseline or BAU emissions, values were adopted from the Climate Equity Reference Project (2016).

Table A.4.1. Unconditional GHG mitigation pledges, 2030

Country	GDP (billions 2015 $)	Population (millions)	Per capita GDP (thousands 2015 $)	BAU emissions ($mtCO_2e$)	Pledged reduction[a] ($mtCO_2e$)	Pledged reduction (% of BAU)
USA	25,926.10	355.82	72.9	6,599.50	2,056.71	31.16
EU	20,900.80	513.80	40.7	4,661.08	1,519.87	32.61
Brazil	2,418.82	228.66	10.6	2,061.00	864.00	41.92
Indonesia	1,813.06	295.48	6.1	2,881.00	835.49	29.00
China	23,263.90	1,413.85	16.5	19,721.40	468.72	2.38
Canada	2,150.55	40.31	53.4	975.00	450.00	46.15
South Korea	2,073.34	52.98	39.1	850.60	314.72	37.00
Japan	4,752.85	120.16	39.6	1,042.00	310.00	22.93
Mexico	1,865.67	148.13	12.6	1,110.00	277.50	25.00
Ethiopia	175.18	138.30	1.3	400.00	256.00	64.00
Australia	2000.02	27.91	71.7	646.43	238.48	36.89
Kazakhstan	336.29	19.42	17.3	461.00	211.53	45.89
Nigeria	949.51	262.60	3.6	900.00	180.00	20.00
Saudi Arabia	1,157.89	39.13	29.6	844.00	129.98	15.40
Thailand	613.05	68.25	9.0	555.00	111.00	20.00
Argentina	780.15	49.37	15.8	670.00	100.50	15.00
Gabon	26.61	2.32	11.5	192.50	96.25	50.00
Angola	207.34	39.35	5.3	193.25	67.64	35.00
Colombia	555.41	53.18	10.4	335.00	67.00	20.00
Vietnam	474.22	99.32	4.8	787.40	62.99	8.00
Peru	362.43	36.86	9.8	298.30	59.66	20.00
Zambia	55.46	25.31	2.2	171.61	42.90	25.00
Paraguay	42.88	7.85	5.5	416.00	41.60	10.00
Iran[b]	857.11	88.53	9.7	1,016.54	40.66	4.00
Switzerland	842.37	9.39	89.7	56.84	29.82	52.31
Azerbaijan	82.62	10.43	7.9	106.81	24.37	22.82
Israel	514.55	10.12	50.8	105.50	23.85	22.61
Morocco	163.99	39.79	4.1	171.00	22.23	13.00
Algeria	267.27	48.27	5.5	310.63	21.74	7.00
Ecuador	164.37	19.56	8.4	158.22	20.40	12.89
Trinidad and Tobago	30.81	1.37	22.5	62.00	18.60	30.00

(continued)

Table A.4.1. Continued

Country	GDP (billions 2015 $)	Population (millions)	Per capita GDP (thousands 2015 $)	BAU emissions (mtCO_2e)	Pledged reduction[a] (mtCO_2e)	Pledged reduction (% of BAU)
Tanzania	124.05	82.93	1.5	153.00	15.30	10.00
Bangladesh	474.78	186.46	2.5	234.00	11.70	5.00
Ghana	76.24	36.87	2.1	74.00	11.09	15.00
Ivory Coast	55.91	32.14	1.7	34.30	9.59	28.00
Burkina Faso	22.32	27.24	0.8	118.30	7.81	6.60
Lebanon	69.99	4.65	15.1	44.00	6.60	15.00
Guatemala	108.52	21.42	5.1	53.90	6.03	11.20
Tunisia	82.40	12.33	6.7	68.20	6.00	8.80
Norway	505.88	5.84	86.6	30.92	5.92	19.10
Georgia	20.20	4.13	4.9	38.40	5.76	15.00
Macedonia	16.84	2.08	8.1	17.66	5.23	29.60
Chad	17.71	21.95	0.8	28.70	5.22	18.20
Sri Lanka	179.05	22.24	8.1	63.25	4.43	7.00
Togo	7.36	10.49	0.7	38.90	4.33	11.14
Serbia	61.74	6.39	9.7	5.57	3.61	5.47
Niger	15.67	35.97	0.4	96.50	3.38	3.50
Eritrea[b]	4.15	7.31	0.6	6.30	2.47	39.20
Senegal	24.60	22.80	1.1	46.42	2.32	5.00
Burundi	7.63	17.36	0.4	74.49	2.23	3.00
Namibia	18.58	3.27	5.7	22.70	2.02	8.90
Oman	100.94	5.24	19.3	90.50	1.81	2.00
Djibouti	3.26	1.05	3.1	4.50	1.79	40.00
Montenegro	0.16	0.61	0.3	4.50	1.57	34.93
Jamaica	21.84	2.91	7.5	14.49	1.13	7.80
Haiti	14.56	12.58	1.2	20.50	1.03	5.00
Bosnia and Herzegovina	26.28	3.58	7.3	40.90	0.82	2.00
Benin	13.53	15.59	0.9	22.50	0.79	3.50
Jordan	59.43	8.03	7.4	51.00	0.77	1.50
Solomon Islands	1.83	0.76	2.4	2.46	0.74	30.00
Albania	18.34	3.05	6.0	6.20	0.71	11.50
Kyrgyzstan	10.26	7.15	1.4	4.50	0.57	12.62
Mauritania[b]	13.37	5.67	2.4	18.80	0.50	2.68
Yemen	81.19	36.34	2.2	43.80	0.44	1.00
Fiji	5.89	0.94	6.3	3.50	0.35	10.00
Gambia	1.79	3.11	0.6	3.90	0.33	8.47

Maldives	7.36	0.49	15.0	3.30	0.33	10.00
Lesotho	4.45	2.49	1.8	2.90	0.29	10.00
Andorra[b]	7.71	0.07	110.1	0.53	0.20	37.00
Seychelles	2.81	0.10	28.1	0.70	0.19	29.00
Saint Vincent and the Grenadines	1.07	0.11	9.7	0.70	0.15	22.00
Dominica	0.64	0.08	8.0	0.22	0.13	58.65
Liechtenstein[b]	11.58	0.04	289.5	0.25	0.11	45.11
Marshall Islands	0.26	0.06	4.3	0.20	0.08	43.47
Monaco[c]	7.06	0.04	176.5	0.11	0.06	50.18
San Marino[c]	1.85	0.03	61.7	0.25	0.04	17.04
Cook Islands[c,d]	0.31	0.02	15.5	0.14	0.03	23.52
Micronesia	0.34	0.12	2.8	0.20	0.02	16.92
Kiribati	0.20	0.14	1.4	0.10	0.02	12.80
India	6,636.33	1,527.66	4.3	5,695.19	−1,413.07	−24.81
Armenia	16.23	2.99	5.4	18.09	−9.13	−101.82
Chile	453.01	20.25	22.4	144.29	−35.66	−24.71
Singapore	474.19	6.39	74.2	33.26	−38.32	−115.23
Tajikistan	12.52	11.10	1.1	21.65	−3.27	−15.12
Uruguay	89.40	3.60	24.8	26.40	−1.03	−3.90
Moldova	11.29	3.22	3.5	9.95	−2.99	−30.03
Belarus	61.54	8.62	7.1	95.00	−5.19	−5.46
Malaysia	541.22	36.11	15.0	636.24	−36.19	−5.69
Ukraine	152.15	40.26	3.8	427.34	−97.42	−22.80
Russia	1,827.71	133.99	13.6	1,864.73	−872.57	−46.79
Total	108,412.13	6,734.22	16.1	59,322.91	9,100.23	15.34

[a] A few countries made pledges such that their goal is above their projected 2030 BAU emissions level, resulting in a "negative" emissions reduction. For our analyses, we constrain these countries to zero emissions change from BAU.
[b] These countries did not have GDP data in the World Bank dataset used to derive GDP deflators from 2010 dollars to 2015 dollars. Thus, we leave them in 2010 dollars and note that the difference for these few countries would not meaningfully affect our analysis.
[c] These countries did not have GDP data in the USDA ERS 2030 projections dataset. Thus, we instead use their current GDP levels (2014) according to the United Nations data (UN, 2016), and note that any difference between current and 2030 GDPs would be relatively small enough as to not significantly affect our analysis.
[d] This country did not have population data in the US Census Bureau 2030 projections dataset. Thus, we instead use their current population level (2016) according to the United Nations data (UN, 2016), and note that any difference between current and 2030 population would be relatively small enough as to not significantly affect our analysis.

References

Anderson, S. and J. Sallee. 2011. "Using Loopholes to Reveal the Marginal Cost of Regulation: The Case of Fuel-Economy Standards." *American Economic Review* 101: 1375–409.

Andresen, S. 2015. "International Climate Negotiations: Top-Down, Bottom-Up or a Combination of Both?" *The International Spectator* 50 (1): 15–30.

Atkinson, A. 1970. "On the Measurement of Inequality." *Journal of Economic Theory* 2: 244–63.

Averchenkova, A., and S. Bassi. 2016. "Beyond the Targets: Assessing the Political Credibility of Pledges for the Paris Agreement." Grantham Research Institute on Climate Change and the Environment, London.

Averchenkova, A., N. Stern, and D. Zhenghelis. 2014. "Taming the Beasts of 'Burden-sharing': An Analysis of Equitable Mitigation Actions and Approaches to 2030 Mitigation Pledges." Grantham Research Institute on Climate Change and the Environment, London.

Bento, A. M., R. Kanbur, B. and Leard. 2015. "Designing Markets for Carbon Offsets with Distributional Constraints." *Journal of Environmental Economics and Management* 70: 51–71.

Braun, D. 1988. "Multiple Measurements of U.S. Income Inequality." *Review of Economics and Statistics* 70 (3): 398–405.

Capros, P., D. Van Regemorter, L. Paroussos, P. Karkatsoulis, T. Revesz, C. Fragkiadakis, S. Tsani, I. Charalampidis (authors); M. Perry, J. Abrell, J. C. Ciscar, J. Pycroft, B. Saveyn (eds.). 2013. "GEM-E3 Model Documentation." JRC Scientific and Technical Reports. EUR 26034 EN.

Clarke-Sather, A., J. Du, Q. Wang, J. Zeng, and Y. Li. 2011. "Carbon Inequality at the Sub-National Scale: A Case Study of Provincial-Level Inequality in CO2 Emissions in China 1997–2007." *Energy Policy* 39: 5420–8.

Climate Equity Reference Project. 2016. "The Climate Equity Reference Calculator." Stockholm Environment Institute. <http://calculator.climateequityreference.org/>.

Du, G., C. Sun, and Z. Fang. 2015. "Evaluating the Atkinson Index of Household Energy Consumption in China." *Renewable and Sustainable Energy Reviews* 51: 1080–7.

Ellerman, A. D., and A. Decaux. 1998. "Analysis of Post-Kyoto CO2 Emissions Trading Using Marginal Abatement Curves." Report No. 40, MIT, Cambridge, MA.

Fair Shares. 2015. "A Civil Society Equity Review of INDCs." Report. <http://civilsocietyreview.org/wp-conte>.

Green Climate Fund. 2016. "Resources Mobilized." <http://www.greenclimate.fund/partners/contributors/resources-mobilized>.

Groot, L. 2010. "Carbon Lorenz Curves." *Resource and Energy Economics* 32: 45–64.

Hedenus, F., and C. Azar. 2005. "Estimates of Trends in Global Income and Resource Inequalities." *Ecological Economics* 55: 351–64.

IPCC (Intergovernmental Panel on Climate Change). 1995. *IPCC Second Assessment Report: Climate Change 1995*. New York: Oxford University Press.

IPCC (Intergovernmental Panel on Climate Change). 1996. *Second Assessment Report: Climate Change 1995. Economic and Social Dimensions*. Cambridge: Cambridge University Press.

IPCC (Intergovernmental Panel on Climate Change). 2014. *Fifth Assessment Report* (AR5). Geneva: Intergovernmental Panel on Climate Change.

Kesicki, F. 2011. "Marginal Abatement Cost Curve for Policy Making: Expert-based vs. Model-Derived Curves." UCL Energy Institute, University College London, United Kingdom. <http://www.homepages.ucl.ac.uk/~ucft347/Kesicki_MACC.pdf>.

Kesicki, F., and P. Ekins. 2012. "Marginal Abatement Cost Curves: A Call for Caution." *Climate Policy* 12 (2): 219–36.

Kverndokk, S., and A. Rose. 2008. "Equity and Justice in Climate Change Policy." *International Review of Environmental and Resource Economics* 2 (2): 135–76.

Live Population. 2016. "Population Projections." <http://www.livepopulation.com/population-projections/world-2030.html>.

McKinsey & Company. 2013. "Greenhouse Gas Abatement Cost Curves." <http://www.mckinsey.com/business-functions/sustainability-and-resource-productivity/our-insights/greenhouse-gas-abatement-cost-curves>.

Morris, J., S. Paltsev, and J. Reilly. 2012. "Marginal Abatement Costs and Marginal Welfare Costs for Greenhouse Gas Emissions Reductions: Results from the EPPA Model." *Environmental Modeling and Assessment* 17 (4): 325–36.

Nordhaus, W., and P. Sztorc. 2013. "DICE 2013R: Introduction and User's Manual." <http://www.econ.yale.edu/~nordhaus/homepage/documents/DICE_Manual_103113r2.pdf>.

Rayner, S. 2010. "How to Eat an Elephant: A Bottom-Up Approach to Climate Policy." *Climate Policy* 10 (6): 615–21.

Rose, A. 1992. "Equity Considerations of Tradable Carbon Emission Entitlements." In S. Barrett et al. (eds.), *Combating Global Warming: Study on a Global System of Tradable Carbon Emission Entitlements*. New York: UN Conference on Trade and Development.

Rose, A., and B. Stevens. 1993, "The Efficiency and Equity of Marketable Permits for CO2 Emissions." *Resource and Energy Economics* 15 (1): 117–46.

Rose, A., B. Stevens, J. Edmonds, and M. Wise. 1998. "International Equity and Differentiation in Global Warming Policy." *Environmental and Resource Economics* 12 (1): 25–51.

Rose, A. and Wei, D. 2008. "Greenhouse Gas Emissions Trading Among Pacific Rim Countries: An Analysis of Policies to Bring Developing Countries to the Bargaining Table." *Energy Policy* 36: 1420–9.

Rose, A., D. Wei, N. Miller, and T. Vandyck. 2017. "Equity, Emissions Allowance Trading and the Paris Agreement on Climate Change." *Economics of Disasters and Climate Change* 1 (3): 203–32.

Rose, A., D. Wei, J. Wennberg, and T. Peterson. 2009. "Climate Change Policy Formation in Michigan: The Case for Integrated Regional Policies." *International Regional Science Review* 32 (4): 445–65.

SCAG. 2012. "Microeconomic and Macroeconomic Impact Analysis of Greenhouse Gas Mitigation Policy Options for the Southern California Climate and Economic

Development Project (CEDP)." Final Report by the Center for Climate Strategies (CCS) prepared for the Southern California Association of Governments and the Project Stakeholder Committee of the CEDP. <http://sustain.scag.ca.gov/Sustainability%20 Portal%20Document%20Library/12-30-12_SCAG_CEDP_Final_Report.pdf>.

Sweeney, J. 2016. *Energy Efficiency: Building a Clean, Secure Economy*. Stanford, CA: Hoover Institution Press.

Teng, F., J. He, X. Pan, and C. Zhang. 2011. "Metric of Carbon Equality: Carbon Gini Index Based on Historical Cumulative Emission Per Capita." *Advances in Climate Change Research* 2 (3): 134–40.

UN (United Nations). (2016). "Country Profiles." United Nations Statistics Division. <http://data.un.org/CountryProfile.aspx>. Accessed 22 February 2016.

UNFCCC. 2015. "COP21 Final Draft." <http://unfccc.int/resource/docs/2015/cop21/eng/l09.pdf>.

UNFCCC. 2016. "Submitted INDCs." <http://unfccc.int/focus/indc_portal/items/8766.php>.

US Census Bureau (2015). "Projected Population and Growth Rates in Population for Baseline Countries/Regions 2010–2030." International Data Base. <http://www.census.gov/ipc/www/idb/>.

US Department of Agriculture. 2015. "Projected Real GDP for Baseline Countries." <http://www.ers.usda.gov/datafiles/International_Macroeconomic_Data/Baseline_Data_Files/ProjectedRealGDPValues.xls>.

Vandyck, T., K. Keramidas, B. Saveyn, A. Kitous, and Z. Vrontisi. 2016. "A Global Stocktake of the Paris Pledges: Implications for Energy Systems and Economy." *Global Environmental Change* 41: 46–63.

Victor, D., J. House, and S. Joy. 2005. "A Madisonian Approach to Climate Policy." *Science* 309 (5742): 1820–21.

Wei, D., and A. Rose. 2014. "Macroeconomic Impacts of the California Global Warming Solutions Act on the Southern California Economy." *Economics of Energy and Environmental Policy* 3 (2): 101–18.

World Bank. 2010. "Generating the Funding Needed for Mitigation and Adaptation." In *World Development Report: Development and Climate Change*. Washington, DC: The World Bank. <http://siteresources.worldbank.org/INTWDR2010/Resources/5287678-1226014527953/WDR10-Full-Text.pdf>.

5

Climate Change and Inequity: How to Think about Inequities in Different Dimensions

Nicole Hassoun and Anders Herlitz

5.1. Introduction

Recent attempts to mitigate climate change, and its devastating effects, focus on the importance of reducing greenhouse gas emissions, but also raise complex equity concerns that are difficult to resolve. Climate policies—e.g. UN-REDD programs; carbon capture, utilization, and storage; and solar radiation management—have distributional consequences, and ensuring a just distribution of benefits and burdens requires taking numerous equity concerns into account (United Nations, 2016). Perhaps because the equity landscape is so complex, those concerned about it make little headway in international negotiations (Maljean-Dubois, 2016; Savaresi, 2016; Morgan and Waskow, 2014; Jamieson, 2014; Light, 2012). The Paris Agreement, for example, largely consists of country pledges and international review systems that we believe fail to give due weight to equity. Countries and other organizations that, like Bolivia in Paris 2015, attempt to place equity issues on the table in international negotiations find little support for their views (Maljean-Dubois, 2016). Meanwhile, research on climate justice underlines climate policies' potentially huge equity consequences (cf. Adler and Treich, 2015; Gardiner, 2006; Shue, 2014). Many argue, for instance, that certain countries should take on larger burdens since they bear greater responsibility for climate change and have benefited more from past greenhouse gas emissions (Pogge, 2002; Ngwadla, 2014).[1] Others argue that ability to pay provides a fair ground for the distribution of burdens, and yet others argue that we should not burden individuals in poverty (Shue, 2014; Caney, 2012). We recognize both the

[1] For criticism, see Karnein (2017).

importance of reducing greenhouse gas emissions and of shifting to low-carbon societies. Moreover, we realize that, sometimes, pragmatic considerations outweigh other things that matter. Still, we believe that equity should take center stage in forming climate policy. So, we propose a simple framework for thinking about equity that can help put it back on the table by assisting stakeholders in evaluating different policies' consequences.

Our model facilitates thinking about three pressing, and sometimes neglected, equity dimensions when evaluating climate policies: distributions of goods across countries, distributions of goods within countries, and distributions of goods between individuals in the world. International negotiations often focus primarily on the former, but we think it is essential to incorporate the latter dimensions as well.[2] In this chapter, we present a model that allows researchers and policymakers to take into account not only how countries compare, but also other salient equity considerations that relate to distributions between individuals globally and within countries. The model will help researchers develop social indicators and indices to inform policymakers about how alternative policies and agreements address equity. As such, the model makes it easier to observe and to address equity considerations in policy development and international negotiations.[3]

This model supports the development of a proposed equity reference framework (ERF) that policymakers can use to evaluate policies and international agreements that aim to combat climate change (BASIC experts, 2011; Holz et al., 2017; Ngwadla, 2014). An ERF would operationalize key normative principles such as equality, sustainable development, and ability to pay, and help stakeholders visualize and measure how well policies fulfill these principles. Different stakeholders can use the framework at different stages of the policymaking process (e.g. they can be used by individuals taking part in national public debates, by political parties and interest groups at national and international levels, and by country representatives in international negotiations). To promote equity, some researchers even provide websites with equity "calculators" to help set targets and evaluate performance in light of different equity considerations.[4] We propose a model for expanding these efforts to outline a more relevant ERF.

Our framework can incorporate the most salient perspectives in climate justice and climate ethics as these fall into the three categories (dimensions)

[2] Although some ethicists are primarily concerned about how individuals within states fare (as opposed to individuals irrespective of location), even such "statists" do not privilege equity between states for its own sake (Blake, 2009; Miller, 2010).

[3] Obviously, this does not solve the problem of how to motive countries to care about equity, but visualizing what we need to do is a first step toward positive change.

[4] See, for instance, the Stockholm Environment Institute "Climate Equity Reference Calculator", <https://www.sei-international.org/equity-calculator>. Accessed May 2018.

suggested above (cf. Adler and Treich, 2015; Boran, 2014; Budolfson et al., 2017; Broome, 2012; Caney, 2012; Gardiner, 2006; Firebaugh, 1999; Moellendorf, 2012, 2015; Shue, 1999, 2014). Much of the literature recognizes that different countries ought to take on different burdens. Typically, approaches of this kind focus on how countries have contributed to climate change in the past and what this entails in terms of responsibility to mitigate the problem (Ngwadla, 2014; Shue, 1999). This relates to *distribution of goods across countries*.[5] Others focus on fairness in distributing burdens and benefits across individuals in the world (cf. Adler and Treich, 2015; Broome, 2012; Caney, 2012; Moellendorf, 2012, 2015). This relates to *distribution of goods between individuals in the world*. Finally, some worry that the transition from high-carbon societies to low-carbon societies creates inequalities in these societies (Morgan and Waskow, 2014). This relates to *distribution of goods within countries*. Our model provides a way to think about these three equity dimensions at the same time, and thereby enables operationalization of normative considerations in all of these three dimensions.

To present our model, we start with some simplifying assumptions that researchers can modify as they operationalize it. First, we stay neutral on what resources, welfare, capabilities etc. matter to individuals (cf. Arneson, 1999; Dworkin, 2000; Nussbaum, 2000) and refer to them simply as *benefits*.[6] We consider cases where the benefits are equally valuable for each individual and there is no diminishing rate of return (where a unit of benefit is equally valuable for someone regardless of how many benefits she has). We simply sum individuals' benefits within each country to measure the benefits accruing to the country. Second, we consider *more equal* distributions of benefits between and across individuals and countries to be more desirable. This is plausible in the context of distributions between individuals in the world and in the context of distributions within countries. It is less plausible when one speaks of distributions across countries, partly because countries have different sizes, but in the climate change context also because one might want to consider responsibility and ability to pay. Importantly, one can easily replace these equality benchmarks with different benchmarks. For example, a benchmark that considers country size, a benchmark that reflects responsibility and ability to pay, or a benchmark that reflects equal ability for sustainable development, could replace

[5] We explain below how these three dimensions allow us to capture many pressing equity concerns in the literature, including responsibility for past emissions, ability to pay, and sustainable development.

[6] This way of staying neutral on what makes lives go well has a long tradition in ethical theory (cf. Arrhenius, 2013; Crisp, 2003; Fleurbaey et al., 2009; Parfit, 1984, Temkin, 2003; Voorhoeve, 2014). Note, also, that our use of a one-dimensional unit of measurement that represents how good a life is and that can be used to measure also how well off a country is essentially mirrors how climate economists often model different approaches (Adler and Treich, 2015; Broome, 2012). However, we call the relevant factors "benefits" and not "consumption" as some benefits may not be things people can consume.

the equality benchmark to evaluate distributions across countries. Similarly, some sufficientarian benchmark that reflects the importance that each individual in the world has their basic needs satisfied could replace the equality benchmark to evaluate distributions across individuals in the world. To illustrate our argument, we consider how just a few contemporaneous individuals fare rather than studying large populations across large time spans. With future research, one can modify our model to accommodate different views and possibilities, e.g. distributions across time in each dimension and other benchmarks than equality. We briefly discuss such modifications at the end of the chapter.

5.2. Different Types of Inequality and Trade-offs Between Them

Climate change actualizes a number of equity problems. How should one distribute climate change mitigation costs between relatively recently industrialized countries (e.g. Botswana, China, and Malaysia) and countries that industrialized in the nineteenth century (e.g. Austria, Sweden, and the United States)? The latter countries have emitted significantly more greenhouse gases than the former. Moreover, inhabitants of the latter countries have benefited from these past emissions. Past activities (e.g. industrial production) that have contributed to climate change partly explain why those in the latter countries are much better off. The transition from a high-carbon society to a low-carbon society also affects different groups in various ways. Generally speaking, individuals working in traditional industries that emit much greenhouse gases (e.g. mining, steel production, trucking) face greater challenges by this transition than individuals who work in newer industries (e.g. academia, advertising, IT). This seems unfair. To explore how to think about climate change equity, we explore how to think about different kinds of inequality.

To illustrate the inequities that relate to climate policy, consider the following case, where Outcome 1 represents the distribution of benefits before one implements an effective climate policy that reduces emissions and increases the total amount of benefits in the world, and Outcome 2 represents the distribution of benefits after implementation:

Outcome 1				
Individual:	x	y	z	Total
Country A	2	2	3	7
Country B	2	3	3	8
Outcome 2				
Individual:	x	y	z	Total
Country A	1	4	2	7
Country B	2	5	5	12

The climate policy increases the total amount of benefits in society (by mitigating the negative effects of climate change), but these benefits are not evenly distributed across countries or individuals. Furthermore, the policy introduces costs that are not evenly distributed: although it increases the total amount of benefits, and although it improves the lives of half of the world's population, the policy harms some individuals. Perhaps workers in traditional industries suffer from stricter environmental regulations, and people who live in (e.g.) coastal areas closer to the equator benefit more from combating climate change than other groups (e.g. those who lose access to farmland when the policy protects forests). In this simple illustration, we see that mitigating climate change can create significant inequalities.

Inequality is complex and has multiple relevant interpretations (cf. Hassoun, 2011; Sen, 1992; Temkin, 1993). Distributions of benefits across different types of entities (e.g. countries, individuals, subgroups with certain characteristics) may be more or less equal. Distributions of benefits within different temporal demarcations (e.g. a year, whole lives, periods of lives) may be more or less equal (McKerlie, 1989). Moreover, one can measure inequality in different ways (e.g. total sum of differences from the mean, total span from the worst off to the best off) (Temkin, 1993). Three distributional considerations are especially pressing when considering climate change policy: distribution across countries, distribution within countries, and distribution across individuals in the world. (One must also consider intergenerational inequality, but we return to this in Section 5.4.)

In this section, we illustrate how concern for equality between these three different entities might have different implications for policy evaluation. More precisely, we consider:

International equality: equal distribution of benefits between countries.
Intra-national equality: equal distribution of benefits between individuals within countries.
World equality: equal distributions of benefits between individuals within the world, irrespective of location.

We focus on these kinds of inequity because they are salient in the literature on climate justice. Many international negotiations and theories that refer to responsibility for past emissions focus on international equality (Broome, 2012; Ngwadla, 2014; Shue, 2014). Those who worry about the consequences of transitioning from high-carbon to low-carbon societies focus on intra-national equality (Morgan and Waskow, 2014). Climate economics and certain theories of climate ethics, for example theories that ascribe greater value to worse-off consumers, tend to focus on world equality (cf. Adler and Treich, 2015; Broome, 2012; Budolfson et al., 2017; Nordhaus, 2008; Schultz, 1998; Stern, 2007).

To illustrate our model, we use the Gini coefficient to measure inequality. When looking at international inequality, we look at inequality between the total sum of benefits in countries. Note that, for outcomes in which countries differ greatly with respect to population size, this is problematic and it might be better to look at inequality with respect to the average amount of benefits individuals in different countries have, or in other ways take population size differences into account.

We calculate the Gini coefficient by looking at the relative mean difference (Sen and Foster, 1997). Thus, for example, in a simple case with only three individuals (A, B, and C) with different amounts of benefits (1, 5, and 10 respectively) we calculate the Gini coefficient as follows:

Relative amount of benefits: A: 1/16 = 0.0625; B: 5/16 = 0.3125; C: 10/16 = 0.625
Mean relative difference: |0.0625 – 0.3125| + |0.0625 – 0.625| + |0.3125 – 0.625|/3 = 0.375
Gini coefficient: 0.375

Consider, first, how a climate policy might affect both intra-national and international equality. Suppose we implement a policy to combat deforestation that transfers resources from a rich country with little forest to a poorer forested country with much forest in order to protect the forests in the poorer country as part of the UN-REDD program. This sort of transfer might reduce inequality in some respect since it moves resources from richer to poorer countries, but this does not ensure that the program promotes equity in other dimensions. Table 5.1 illustrates what the transfer might entail in terms of the distribution of benefits. Let Outcome 1 be the situation before we implement the policy and let Outcome 2 be the situation after implementation.

As we see, international inequality goes down after we implement the policy, but as an effect of the resource transfer we observe an increase in intra-national inequality. There is no intra-national inequality in Outcome 1, and there is

Table 5.1. Case 1

Outcome 1				
Individual	x	y	z	Total
Country A	1	1	1	3
Country B	4	4	4	12
Outcome 2				
Individual	x	y	z	Total
Country A	1	1	6	8
Country B	2	3	5	10

considerable intra-national inequality in Outcome 2.[7] Although it in some sense decreases inequality (at country level), transferring resources from a richer country to a poorer country might increase the amount of inequality in a different sense. This depends on who loses in the richer country and who wins in the poorer country. In our simple model, only individual 3 in the poorer country is better off after we implement the policy and only individuals 1 and 2 in the richer country are worse off. In effect, the policy creates inequalities.

In light of this, one might consider whether international agreements do a better job of safeguarding equality if they contain provisions at the country level to require widespread sharing of the benefits gained and costs imposed by international treaties.

Yet, climate policies may have other equity effects. Climate policies might have detrimental effects in terms of world inequality. To see this, consider a more ambitious climate policy that transfers significantly more resources but that still creates some intra-national inequality (Table 5.2)

Table 5.2. Case 2

Outcome 1				
Individual	x	y	z	Total
Country A	1	1	1	3
Country B	4	4	4	12
Outcome 3				
Individual	x	y	z	Total
Country A	2	3	3	8
Country B	3	3	4	10

Since Outcome 3 contains intra-national inequality, concern for intra-national equity provides some reason to favor Outcome 1 over Outcome 3. Yet, Outcome 1 contains significantly more world inequality than Outcome 3. Using the Gini coefficient to measure world inequality we can observe the following amounts of inequality in the two outcomes: Outcome 1=0.12; Outcome 3=0.093. So, Outcome 3 is better than Outcome 1 in this respect. Thus, although this policy creates some intra-national inequality, it reduces world inequality significantly. In light of this, perhaps we need international agreements to include provisions to also reduce world inequality.

Still, policies could also be objectionable because they exacerbate international inequality. Concern for international and world equity can entail different evaluations. Suppose policymakers modified the details of the UN-REDD program to produce Outcome 4 (Table 5.3).

[7] The intra-national Gini coefficient for both countries is 0 in Outcome 1. The intra-national Gini coefficient for Country A in Outcome 2 is 0.467. The intra-national Gini coefficient for Country B in Outcome 2 is 0.2.

Table 5.3. Case 3

Outcome 1				
Individual	x	y	z	Total
Country A	1	1	1	3
Country B	4	4	4	12
Outcome 4				
Individual	x	y	z	Total
Country A	1	2	6	9
Country B	1	2	6	9

In Outcome 4, the distribution of benefits across countries is equal whereas Outcome 1 contains significant international inequality. Thus, Outcome 4 seems better than Outcome 1. Yet, Outcome 4 contains significantly more world inequality than Outcome 1 (0.01422 vs. 0.37). This policy creates perfect equality at the country level, but it does so at the expense of both intra-national and world equality.

As this discussion makes clear, concern for different types of equality can rank outcomes differently. This partly explains the difficulty of considering equity when outlining climate policy. Different equity considerations sometimes point in different directions, and it is not clear how to think about this. One outcome might contain less intra-national equality while a different outcome has less world equality. Moreover, the best outcome in terms of either intra-national equality or world equality may not be best in terms of international equality. This is not always the case, and most notably, perhaps, it is not the case for ideal outcomes when population sizes are equal. An equal distribution across all individuals in the world is by definition an equal intra-national distribution, and in case all relevant groups (e.g. countries) contain the same number of individuals it also follows analytically that the distribution across these groups is equal. International agreements might aim to reduce all kinds of inequality. However, it may be difficult to achieve an ideal result.

When addressing climate justice and different climate policies, we must also engage with outcome pairs like those in Table 5.4. Here, we let Country A be 1/3 larger than Country B (Individual 4). Which outcome is most equitable? This is not obvious, and different equity concerns provide different answers to the question. Outcome 5 has less international inequality (0.033 vs. 0.063), but there is less world inequality in Outcome 6 (0.202 vs. 0.286). Moreover, there is less intra-national inequality in Outcome 5 than in Outcome 6 (for Country A: 0.321 in Outcome 5 vs. 0.107 in Outcome 6; for Country B: 0.083 in Outcome 5 vs. 0.067 in Outcome 6). If we had to choose between different ways of modifying an international agreement that would each result in one of these outcomes, which should we choose? To answer

Table 5.4. Case 4

Outcome 5					
Individual	w	x	y	z	Total
Country A	1	1	1	4	7
Country B	2	3	3	–	8
Outcome 6					
Individual	w	x	y	z	Total
Country A	1	2	2	2	7
Country B	3	3	4	–	10

this question, one needs some way of evaluating the relative importance of the different equity dimensions. In Section 5.4, we present a model that facilitates thinking about this issue.

5.3. Equality and Other Things that Matter

Equity is important, yet other things also matter when evaluating different distributions, especially when we focus on policies that aim at mitigating climate change. In addition to equity (at least understood as equality or distribution of benefits and burdens), we generally care about net benefits. This is particularly true when we consider climate change policy. Considering that, for example, deforestation has been estimated to cause more than 15 percent of global emissions, the welfare effects of combating climate change with deforestation policies may swamp concern for equity (Scientific American, 2017).

To see why focusing only on equality is problematic, consider a distribution in which *everyone* is equal but extremely badly off, as presented in Table 5.5 (cf. Parfit, 1997; Temkin, 1993). This outcome contains no international, intra-national, or world inequality. This does not mean that it is better than the outcomes discussed in Section 5.2. A significant number of people are significantly worse off in Outcome 7 compared to each of the outcomes discussed in Section 5.2. This simple illustration reveals that we must weigh concern for equality against other normative considerations. In this section, we discuss the trade-off between the total sum of benefits in an outcome and the amount of equality it contains.

First, social planners must often make trade-offs between the total amount of benefits that are distributed and inequality. Suppose that one compares two climate policies. People do slightly less well under the first policy because it

Table 5.5. Case 5

Outcome 7				
Individual	x	y	z	Total
Country A	1	1	1	3
Country B	1	1	1	3

Table 5.6. Case 6

Outcome 3				
Individual	x	y	z	Total
Country A	2	2	3	7
Country B	2	2	4	8
Outcome 8				
Individual	x	y	z	Total
Country A	1	3	4	8
Country B	2	3	6	11

does less to combat climate change, but the second contains much more inequality. In the second situation, most people are better off, but certain groups are worse off in each country. In the case of an anti-deforestation program, this might happen if the policy requires removing indigenous people from their lands. Suppose this policy results in Outcome 8 (Table 5.6).

On all the accounts of inequality already discussed, Outcome 8 contains more inequality than Outcome 3 (International: 0.1578 vs. 0.3334; Intranational: Country A 0.1666 vs. 0.0952; Country B 0.2424 vs. 0.1666; World: 0.1052 vs. 0.0578). Yet, only one individual is worse off in Outcome 8 compared to Outcome 3, and four individuals (≈ 67 percent of the total population) are better off in Outcome 8 compared to Outcome 3. The increase in the total amount of benefits in Outcome 8 makes this possible. Most egalitarians would accept that equality is not all that matters and, in cases such as this, one must weigh equality against the value of increasing the total sum of benefits.

One might combine consideration for the total amount of benefits and fair distribution by saying that equality lacks value, but holding that it is better to have more equality and that sometimes a smaller amount of benefit is preferable if it is better for the worse off (Adler, 2012). This view is *prioritarian* (Parfit, 1997). According to prioritarianism, benefits to the worse off matter more. On this view, one evaluates outcomes with respect to aggregate benefits, but ascribes more value to benefits to the worse off. Since benefits always have some positive weight on this view, one cannot improve just outcomes by making people worse off. A prioritarian may be particularly concerned about the least well-off people in all countries and strongly prefer an international agreement that results in Outcome 3 to one that results in Outcome 8 even though each country has fewer benefits.

One can present prioritarian views in each of the dimensions discussed earlier. One might be an intra-national prioritarian, i.e. say that how good an outcome is depends on how much priority-weighted benefits there are in different countries. One might be an international prioritarian, i.e. say that the goodness of an outcome depends on how much priority-weighted benefits different countries have. Alternatively, one might be a world prioritarian, i.e. say that an outcome's quality depends on how much priority-weighted benefits there are in the world. One might of course accept all of these views.

In what follows, we remain focused on different types of equality rather than priority-weighted benefits. Yet, note that those who do not think we can improve an outcome by making people worse off can replace the egalitarian perspective with a prioritarian one (or many others).

5.4. A Framework for Measuring Inequality

It is hard to account for equity when one evaluates policies that affect international, intra-national, and world equality partly because different considerations sometimes rank different policies differently. In this section, we present a model that makes it simpler to think about equity. Our proposed *inequality maps* can accommodate intra-national, international, and world inequality (Figure 5.1).

Suppose the dots (vertices) denote people and the lines that connect the dots (edges) indicate a comparison. We compare each segment (vertex) with each person's benefits to each segment (vertex) with every other person's benefits. We capture intra-national equality in the comparisons at the top and bottom of the diagram. World inequality requires comparisons between all vertices. We capture international inequality by introducing one more feature to these inequality maps: sets of segments within which we compare intra-national inequalities—the circles around the dots on the upper and lower lines.

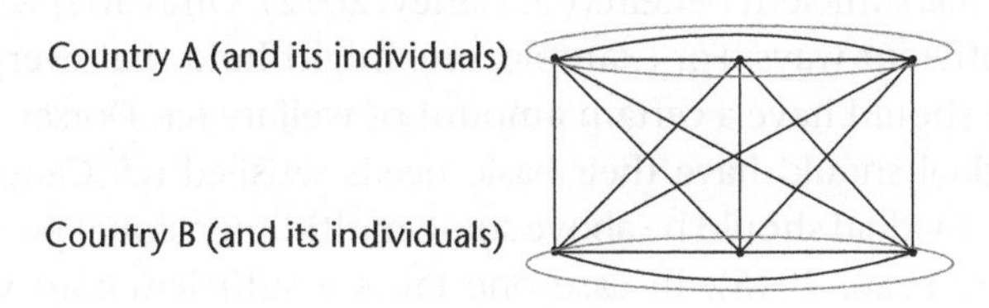

Figure 5.1. Case 7

Table 5.7. Case 8

Individuals	1	2	Total
Country A:	1	2	3
Country B:	3	4	7

Suppose we want to evaluate the simple case represented in Table 5.7. Assuming we continue to measure inequality with the Gini coefficient, world inequality is 0.1666. Intra-national inequality in Country A is 0.3334. Intra-national inequality in Country B is 0.2858. International inequality is 0.4. So, the *total sum* of inequality in this situation is 0.1666 + 0.3334 + 0.2858 + 0.4 = 1.3050. Now some of these inequalities may not matter at all and some might matter more than others. We can capture this by introducing a (for example, linear) weighting function on each kind of inequality. So let t = total inequality, w = world inequality, ic_A = intra-national inequality in Country A, ic_B = intra-national inequality in Country B, and n = international inequality. $t = aw + bic_A + cic_B + dn$ where a, b, c, and d represent the weights given to each kind of inequality. If one cares only about world inequality, one would set b, c, and d equal to zero. If one cares twice as much about intra-national and international inequality as world inequality but only cares about intra-national inequality between citizens of Country A, one might set $a = 1$, $b = 2$, $c = 0$, and $d = 2$. In principle, one could also abandon the simplifying additivity assumption, of course, and allow for different types of inequality to affect the disvalue and weight of other types of inequality. Inequality maps can in this way incorporate different views about the value of different kinds of inequality. Policymakers can use the output of inequality assessments like these to establish equity weights for social welfare functions that consider distributions' fairness.

Policymakers and participants in climate policy negotiations can also use the model to develop equity indices for an ERF to analyze different climate policies. These might take the general form introduced above: $t = aw + bic_1 + cic_2 + dn$, but one can adjust the formula depending on the specific view one takes toward the different equity dimensions. One might replace world inequality, for example, with a sufficientarian view according to which the ideal outcome is not perfect equality but an outcome in which each individual in the world has sufficient benefits (cf. Caney, 2012). One can specify this type of view in different ways. For example, one might hold that every individual in the world should have a certain amount of welfare (cf. Dorsey, 2012), that every individual should have their basic needs satisfied (cf. Caney, 2012), or that every individual should be above some multidimensional poverty threshold (cf. Alkire et al., 2015). In case one takes a sufficientarian view toward world distribution, w would be a shortfall value similar to poverty measures, e.g. the Foster-Greer-Thorbecke indices (Foster et al., 1984; Herlitz and Horan, 2016, 2017). Alternatively, one could let *w* represent the prioritarian idea that benefits to the worse off are more desirable (cf. Adler, 2012; Adler and Treich, 2015). In that case, *w* is either an uncapped measure of priority claim satisfaction or a measure of aggregate shortfall from some unattainable benchmark. Likewise, one can replace international inequality with a view based on responsibility for contributing to climate change and ability to pay if one thinks, for instance, that each country should pay for climate change mitigation according to responsibility and ability (cf. Ngwadla, 2014). Or, one can replace it with the view that each country should have equal access to sustainable development or benefits in some other dimension, or with some combination of these (cf. Shue, 1999). *n* would then be a summary measure of shortfalls from individuated target levels where different targets might reflect things like responsibility to pay, sustainable development, or some combination of equity considerations (cf. Herlitz and Horan, 2017). And in a similar fashion, one can replace intra-national inequality with the view that each individual in each society should have an equal capacity to contribute to society and take part in political life (cf. Anderson, 1999), or that there should be a minimum of status differences in society (cf. O'Neill, 2008), or something else. *ic* might also be some summary measure of shortfalls. How to weight the different dimensions depends on the relative importance one ascribes to them.

Notably, policymakers can also use inequality maps to compare inequalities between populations living at entirely different times, which is particularly interesting when considering climate change policy. So, we might suppose the people in Country A live at times T1–T4 in the cases above and those in Country B live at times T5–T8, and nothing would change in the above analysis. Alternately, one can introduce sets within countries to represent different generations or populations (the smaller circles around subsets of dots on the upper and lower lines) (Figure 5.2).

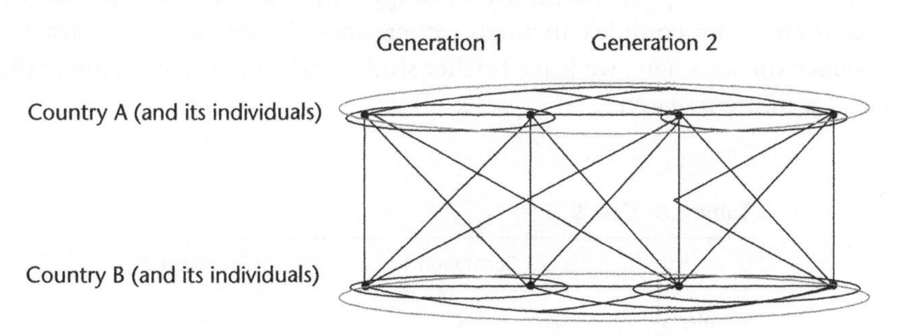

Figure 5.2. Inequality maps comparing countries and generations

So let $g1$ = generation 1 and $g2$ = generation 2, t = total inequality, w = world inequality, ic_A = intra-national inequality in country A, ic_B = intra-national inequality in country B, and n = international inequality. Let a, b, c, d, e, and f represent the weights given to each kind of inequality. Again, if total inequality is a linear function of generational inequalities we can use the weights e and f for generation 1 and 2 respectively and say:

$$t = e(t_{g1}) + f(t_{g2})$$

$$e(t_{g1}) = aw_{g1} + bic_{Ag1} + cic_{Bg1} + dn_{g1}$$

$$f(t_{g2}) = aw_{g2} + bic_{Ag2} + cic_{Bg2} + dn_{g2}$$

Doing so lets one introduce a discount factor by giving different weights to different generations. Suppose, for instance, that one cares only about world inequality in the case below and discounts by 0.05 the amount of benefits people have at each time period after the first. That is, suppose one sets $a = 1$, $b = 0$, $c = 0$, $d = 0$, $e = 1$, and $f = 0.95$ (Table 5.8).

$$t = e(t_{g1}) + f(t_{g2}) = 0.167 + 0.95 = 2.62$$

$$1(t_{g1}) = 1(.167) + 0ic_{Ag1} + 0ic_{Bg1} + 0n_{g1} = 0.167$$

$$0.95(t_{g2}) = 1(0.1) + 0ic_{Ag2} + 0ic_{Bg2} + 0n_{g2} = 0.095$$

Adding a temporal perspective also lets us evaluate inequities from a perspective located in time. We may sometimes have no reason to consider past inequalities. This would, for example, be the case if one takes a completely future-oriented perspective about how to distribute the costs and benefits of different policies for dealing with climate change as in (a) in Figure 5.3. On the other hand, some prefer to look exclusively at the past as in (b) in Figure 5.3. Here are these options where the circle indicates where we are located.

Here we have left out the lines for comparisons that have zero weight. Again, one can drop the additivity assumption in each of the equations or make them more complex in many other ways. As we have only tried to introduce the idea here, we leave further study, and refinement, of inequality maps for future research.

Table 5.8. Case 9

	Generation 1	Generation 2
Country A	2	4
Country B	4	6

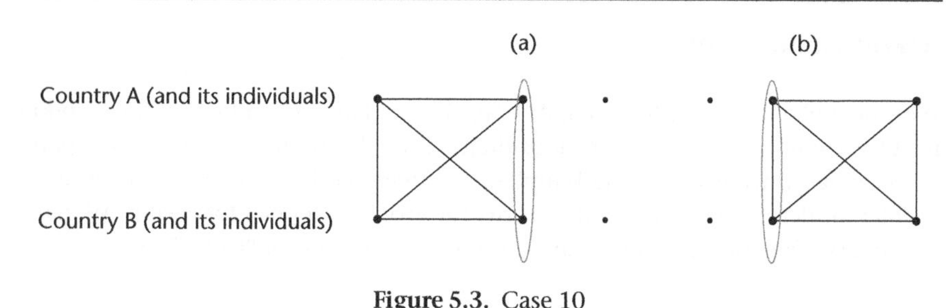

Figure 5.3. Case 10

5.5. Conclusion

How should we evaluate different ways of addressing climate change in light of equity considerations? We need to answer this question to arrive at just climate change policies as different policies—from carbon trading to geoengineering—have different distributional consequences that we should not ignore. Besides the value of creating just outcomes, we need to take the distributional effects of policies into account to ensure sustainability. If we treat countries unfairly in climate change negotiations, they may be unwilling to contribute to solving the problem. If we treat individuals unfairly, their protests may undermine efforts to address climate change, and there are reasons to believe that inequalities undermine sustainability also within countries (cf. Islam, 2015).

This chapter provides a new theoretical framework for evaluating climate change policies. Inequality maps allow one to take into account several different kinds of equity—within and between countries as well as individuals irrespective of country of origin. They can also incorporate concern for some of the other things that matter such as total welfare. Inequality maps can help us capture the different concerns that states and individuals express about climate change policy, especially given (as we have shown) that different kinds of inequality (and other things that matter) do not always go hand in hand: good distributions across countries may, for instance, allow for very bad distributions across individuals. Similarly, concern for distribution and total amount of goods can conflict.

We believe the framework we provide can help us think about the conditions necessary for achieving sustainable development and a just transition to low-carbon societies, the distribution of emissions rights, development aid and capacity building, and technology transfer. It may also help evaluate pathways for empowerment, and different distributions of rights and responsibilities for mitigation and adaptation, as well as compensation for loss and damage (Boran, 2014). We hope that this discussion helps bring equity considerations to the forefront of climate change negotiations.

Acknowledgments

We would like to thank Idil Boran, Avi Appel, Anja Karnein, and Darrel Moellendorf for very useful feedback on a previous draft. Nicole Hassoun is thankful for support from Cornell University and the Templeton Foundation that enabled her to write the chapter. Anders Herlitz is grateful to Forte, the Swedish Council for Health, Working Life and Welfare for providing financial support, grant number 2014–2724.

References

Adler, Matthew. 2012. *Well-Being and Fair Distribution: Beyond Cost–Benefit Analysis*. Oxford: Oxford University Press.

Adler, Matthew, and Nicolas Treich. 2015. "Prioritarianism and Climate Change." *Environmental and Resource Economics* 62 (2): 279–308.

Alkire, Sabina, James Foster, Suman Seth, Maria Emma Santos, and José Manuel Roche. 2015. *Multidimensional Poverty Measurement and Analysis*. Oxford: Oxford University Press.

Anderson, Elizabeth. 1999. "What is the Point of Equality?" *Ethics* 109 (2): 287–337.

Arneson, Richard. 1999. "Human Flourishing Versus Desire Satisfaction." *Social Philosophy & Policy* 16 (1): 113–42.

Arrhenius, Gustaf. 2013. "Egalitarian Concerns and Population Change." In *Inequalities in Health: Concepts, Measures, and Ethics*, edited by Nir Eyal, Samia Hurst, Ole Norheim, and Dan Wikler. Oxford: Oxford University Press.

BASIC experts. 2011. "Equitable Access to Sustainable Development: Contribution to the Body of Scientific Knowledge." BASIC expert group, Beijing, Brasilia, Cape Town, and Mumbai.

Blake, Michael. 2009. *Foreign Policy and Liberal Justice*. Seattle: University of Washington.

Boran, Idil. 2014. "Risk-Sharing: A Normative Framework for Climate Negotiations." *Philosophy and Public Policy Quarterly* 32 (2): 4–13.

Broome, John. 2012. *Climate Matters: Ethics in a Warming World*. New York: W. W. Norton & Company, Inc.

Budolfson, Mark, Francis Dennig, Marc Fleurbaey, Asher Siebert, and Robert H. Socolow. 2017. "The Comparative Importance for Optimal Climate Policy of Discounting, Inequalities and Catastrophes." *Climate Change* 145 (3–4): 481–94.

Caney, Simon. 2012. "Just Emissions." *Philosophy & Public Affairs* 40 (4): 255–300.

Crisp, Roger. 2003. "Equality, Priority, and Compassion." *Ethics* 113 (4): 745–63.

Dorsey, Dale. 2012. *The Basic Minimum: A Welfarist Approach*. Cambridge: Cambridge University Press.

Dworkin, Ronald. 2000. *Sovereign Virtue*. Cambridge, MA: Harvard University Press.

Firebaugh, Glenn. 1999. "Empirics of World Income Inequality." *American Journal of Sociology* 104 (6): 1597–630.

Fleurbaey, Marc, Bertil Tungodden, and Peter Vallentyne. 2009. "On the Possibility of Nonaggregative Priority for the Worst Off." *Social Philosophy and Policy* 26 (1): 258–85.

Foster, James, Joel Greer, and Erik Thorbecke. 1984. "A Class of Decomposable Poverty Measures." *Econometrica* 52 (3): 761–6.
Gardiner, Stephen. 2006. "A Perfect Moral Storm: Climate Change, Intergenerational Ethics and the Problem of Moral Corruption." *Environmental Values* 15 (3): 397–413.
Hassoun, Nicole. 2011. "Free Trade, Poverty, and Inequality." *Journal of Moral Philosophy* 8 (1): 5–44.
Herlitz, Anders, and David Horan. 2016. "Measuring Needs for Priority Setting in Healthcare Planning and Policy." *Social Science and Medicine* 157: 96–102.
Herlitz, Anders, and David Horan. 2017. "A Model and Indicator of Aggregate Need Satisfaction for Capped Objectives and Weighting Schemes for Situations of Scarcity." *Social Indicators Research* 133 (2): 413–30.
Holz, Christian, Sivan Kartha, and Tom Athanasiou. 2017. "Fairly Sharing 1.5: National Fair Shares of a 1.5° C-compliant Global Mitigation Effort." *International Environmental Agreements: Politics, Law and Economics* 8 (1): 117–34.
Islam, S. Nazrul. 2015. "Inequality and Environmental Sustainability." *DESA Working Paper* No. 145. ST/ESA/2015/DWP/145.
Jamieson, Dale. 2014. *Reason in a Dark Time: Why the Struggle Against Climate Change Failed—and What it Means for Our Future.* Oxford: Oxford University Press.
Karnein, Anja. 2017. "Asking Beneficiaries to Pay for Past Pollution." In *Climate Justice and Historical Emissions*, edited by Lukas Meyer and Pranay Sanklecha. Cambridge: Cambridge University Press.
Light, Andrew. 2012. "Climate Ethics for Climate Action." In *Environmental Ethics: What Really Matters? What Really Works?* 2nd edition, edited by David Schmidtz and Elizabeth Willott. New York: Oxford University Press.
Maljean-Dubois, Sandrine. 2016. "The Paris Agreement: A New Step in the Gradual Evolution of Differential Treatment in the Climate Regime?" *Review of European, Comparative & International Environmental Law (RECIEL)* 25 (2): 151–60.
McKerlie, Dennis. 1989. "Equality and Time." *Ethics* 99: 475–91.
Miller, Richard. 2010. *Globalizing Justice: The Ethics of Poverty and Power.* Oxford: Oxford University Press.
Moellendorf, Darrel. 2012. "Climate Change and Global Justice." *Wiley Interdisciplinary Reviews (WIREs) Climate Change* 3 (2): 131–43.
Moellendorf, Darrel. 2015. "Climate Change Justice." *Philosophy Compass* 10 (3): 173–86.
Morgan, Jennifer, and David Waskow. 2014. "A New Look at Climate Equity in the UNFCCC." *Climate Policy* 14 (1): 17–22.
Ngwadla, Xolisa. 2014. "An Operational Framework for Equity in the 2015 Agreement." *Climate Policy* 14 (1): 8–16.
Nordhaus, William. 2008. *A Question of Balance.* New Haven, CT: Yale University Press.
Nussbaum, Martha. 2000. *Women and Human Development: The Capabilities Approach.* Cambridge: Cambridge University Press.
O'Neill, Martin. 2008. "What Should Egalitarians Believe?" *Philosophy & Public Affairs* 36 (2): 119–56.
Parfit, Derek. 1984. *Reasons and Persons.* Oxford: Oxford University Press.
Parfit, Derek. 1997. "Equality and Priority." *Ratio* 10 (3): 202–21.

Pogge, Thomas. 2002. *World Poverty and Human Rights: Cosmopolitan Responsibilities and Reforms*. Cambridge: Polity Press.

Savaresi, Annalisa. 2016. "A Glimpse into the Future of the Climate Regime: Lessons from the REDD+ Architecture." *Review of European, Comparative & International Environmental Law (RECIEL)* 25 (2): 186–96.

Schultz, T. Paul. 1998. "Inequality in the Distribution of Personal Income in the World: How it is Changing and Why." *Journal of Population Economics* 11 (3): 307–44.

Scientific American. 2017. "Deforestation and Its Extreme Effect on Global Warming." <https://www.scientificamerican.com/article/deforestation-and-global-warming/> accessed 26 July 2017.

Sen, Amartya. 1992. *Inequality Reexamined*. Oxford: Clarendon Press.

Sen, Amartya, and James Foster. 1997. *On Economic Inequality*. Oxford: Clarendon Press.

Shue, Henry. 1999. "Global Environment and International Inequality." *International Affairs* 75 (3): 531–45.

Shue, Henry. 2014. *Climate Justice: Vulnerability and Protection*. Oxford: Oxford University Press.

Stern, Nicholas. 2007. *The Economics of Climate Change*. Cambridge: Cambridge University Press.

Temkin, Larry. 1993. *Inequality*. New York: Oxford University Press.

Temkin, Larry. 2003. "Egalitarianism Defended." *Ethics* 113 (4): 764–82.

United Nations. 2016. "Report: Inequalities Exacerbate Climate Impacts on Poor." United Nations, New York. <http://www.un.org/sustainabledevelopment/blog/2016/10/report-inequalities-exacerbate-climate-impacts-on-poor/> accessed 8 July 2017.

Voorhoeve, Alex. 2014. "How Should We Aggregate Competing Claims?" *Ethics* 125 (1): 64–87.

6

Climate Change and Economic Self-Interest

Julie A. Nelson

> The United States or China or Tuvalu (to choose a tiny little economy)—none of them are [participating in COP21] to save the planet. All of these countries are putting their best foot forward because they understand it's good for their economies. And that is the most powerful driving force—the self-interest of every country is what is behind all of these measures. It's not because they want to save the planet.
>
> Christiana Figueres (CBS News, 2015)

> Humans don't have a stronger guiding force than my own self-interest. True for you, true for me, but it's also true at the national level. There is no stronger guiding force than a national interest.
>
> Christiana Figueres, quoted in *The Guardian* (Harvey, 2015)

6.1. Introduction

The climate summit that took place in Paris in late 2015, COP21, was directed by Christiana Figueres, the United Nations climate chief. In many speeches and interviews, she emphasized one point over and over: this summit would reach an agreement because nations are now recognizing that it is in their national self-interest—and, more precisely, in their national *economic* self-interest—to reach a pact. Moving toward cleaner technologies makes a country more "competitive in the global economy," she argued, while also reducing problems at home such as health-damaging levels of pollution (CBS News, 2015).

Let me grant right away that there is some truth to her argument: much of the "low-hanging fruit" of innovations that could serve both climate goals and economic goals remain to be picked. Many things could be done that could serve both national economic self-interest and global sustainability. Many leaders of large companies express their support for climate change policies in terms of the "business case" (by which they mean profit motivations) for action, with references to the advantages it would bring in terms of economic prosperity, competitiveness, and growth (Tabuchi, 2016). To the extent that Figueres's argument prods governments to take some (easy, first) steps away from disastrous current practices and on to a clean energy track, it might be seen as a move in the right direction.

This chapter, however, argues that in a deeper sense—in terms of ethics, human motivation, and how we think about the world we live in—her argument is dangerous. It both draws from, and potentially lends fuel to, views that are deeply problematic from the point of view of climate justice.

6.2. The Power of Economic Self-Interest: Says Who?

Is her assertion about national, and human, behavior true? Is economic self-interest truly the *most powerful* motivator? Is it really the *best* thing to rely on, in our search for a sane and equitable solution to the climate change crisis?

The overwhelming response from the mainstream of my own profession, the economics discipline, would seem to be "yes." After all, rational behavior in the service of economic self-interest is the foundational assumption of our core explanatory—and, we hope, predictive—models. And economists' assumptions have shaped the way many in the profession and beyond see the world around us. Nations should maximize GDP per capita, because GDP represents people's well-being, many have come to believe. The fundamental nature of a business is to maximize profits, nearly everyone now agrees. The image of the rational, self-interested, autonomous actor has also influenced the thinking of some philosophers, and been extended to the level of rational, "self"-interested nation states by scholars of political science and international relations.

Touchy-feely, sentimental stuff such as caring for others or "saving the planet," on the other hand, looks weak, soft, and sissified compared to economists' assumptions. These surely seem to belong to some more personal, tender, feminized sphere. At the most, they might be thought to affect individuals' decision-making in the spheres of family and community. They would seem to have no role in the "tough," macho realms of economics and politics, or within the dog-eat-dog world of competition among nation states. One might fear being thought of as soft and weak, or being laughed out of the room, if one brings up such concerns in a room of "serious men" discussing business or international affairs.

Furthermore, according to neoclassical economic orthodoxy, concerns beyond self-interest are largely unnecessary for the maximization of well-being. In a perfectly competitive, complete system of markets with full information, all factors of production and outputs will be priced appropriately, and efficiency (the only recognized goal) will ensue. Recognizing that these conditions for perfect markets may often not hold—unowned and unpriced environmental resources (such as a stable climate), for example, violate assumptions about information and completeness—some economists have moved on to new theoretical territory. Researchers in the field of "mechanism design," while still maintaining the assumption of self-interest, search for the ways in which an enlightened planner could make people behave in desired ways by creating systems of incentives that align individual self-interest with broader enterprise or social goals. Figueres's appeal to rational self-interest is in this spirit.

Investigation, however, suggests that such an approach is destined for failure. First, it does not work—even in theory. Second, considerable social science research contradicts the claim that self-interest is the strongest motivator.

6.3. Destined for Failure

Is it possible, along mechanism design lines, to create an efficient solution to climate change that relies only on national self-interest? Eric Posner and David Weisbach's 2010 book *Climate Change Justice* addresses exactly this question. They write, "we need to think about how to solve the climate problem in a way that even selfish states would agree to" (2010, p. 138). They look to create strategies that preserve "the principle of International Paretianism: all states must believe themselves better off by their lights as a result of the climate treaty" (p. 6). They dismiss notions of collective responsibility as being contrary to the "standard assumption" of individualism (p. 101). So far, this is quite parallel to Figueres's approach to the Paris talks.

Where does the "justice" aspect promised in their book's title come in? Posner and Weibach's analysis does conclude that "an optimal climate treaty . . . could well require side payments" in which some countries make transfers to others. But rather than these payments being directed *toward* the poor countries that are the most damaged by the long-term greenhouse gas emissions, and coming *from* industrialized countries which are more financially able to cope, the authors see the transfers going "to rich countries like the United States" (p. 86). That is, they envision those countries that suffer the most damage making payments to the countries least affected in order to make mitigation policies serve the self-interest of the rich countries. The use of the word "justice" in their book title appears to be an extreme abuse of language.

Yet even Posner and Weisbach, with all their theoretical hardware and cleverness, are not able to complete their project successfully. Designers of incentive systems want their policies to be "incentive compatible," meaning that, once the systems are place, the parties would abide by them voluntarily. But while nations could in theory be incentivized to sign on to the treaty Posner and Weisbach propose, and would prefer that everyone else abide by the treaty, it would still be in their own national self-interest to cheat once the treaty was signed. Recognizing this, Posner and Weisbach are forced to appeal to the very kind of ethical sensibility they had up to that point been studiously avoiding: "But the obligation to achieve a broad, deep, and enforceable treaty imposes a serious ethical duty on rich and poor nations alike—the obligation to cooperate. In our view, it is unethical for a nation to refuse to join a climate treaty in order to free-ride off of others" (p. 170). Assiduously avoiding ethical considerations in the arena of distributional concerns (rather conveniently, from the point of view of rich nations), the authors still cannot avoid bringing ethical commitments in—even if only as a *deux ex machina* at the end of their story.

Samuel Bowles (2016) points out that that such failures are, in fact, endemic to the whole field of mechanism design. As that field has evolved, it has become apparent that it is not possible, even in theory, to create policies to redress market failures without violating some of the theory's fundamental building blocks. That is, it is impossible to design incentive systems that simultaneously achieve Pareto efficiency, respect private information, allow for any type of preferences, and are incentive compatible (Bowles, 2016, pp. 158, 165). As a result, it is becoming increasingly clear that values beyond self-interest are generally necessary for solving collective problems.

6.4. Evidence on Human Motivations

Many outside the field of economics have wondered why economists adopted the *homo economicus* model of self-interested, autonomous, rational human behavior in the first place. Not only does it fly in the face of centuries of thinking in philosophy and the humanities, it is also contradicted by copious empirical evidence. Those who, as a practical matter, have to deal with the question of motivating people have long recognized the social and emotional complexity of the task (Herzberg, 1987). Recent work in social psychology (Fiske, 2010) and cognitive neuroscience (Adolphs, 2009) adds to the mountain of evidence pointing to the complexity and sociality of human motivations. Moral judgments and moral behavior have been specific areas of intense study (Greene and Haidt, 2002), bringing to light the importance of concerns such as loyalty and sanctity.

Amartya Sen attempted to introduce ethical factors beyond personal preferences into economics in 1977. In his "Rational Fools" essay, he distinguished between "sympathy" and "commitment":

> The former corresponds to the case in which the concern for others directly affects one's own welfare. If the knowledge of torture of others makes you sick, it is a case of sympathy; if it does not make you feel personally worse off, but you think it is wrong and you are ready to do something to stop it, it is a case of commitment....
>
> The characteristic of commitment with which I am most concerned here is the fact that it drives a wedge between personal choice and personal welfare, and much of traditional economic theory relies on the identity of the two.
>
> (Sen, 1977, pp. 326, 329)

Sympathy may, indeed, be too weak a motivator for the purposes of climate justice, depending as it does on self-interest, subjective preferences, and perhaps highly variable emotional responses. Commitment is more promising, as it goes beyond self-interest to the question of our moral commitments and who we understand ourselves to be.

While Sen's distinction failed to have a lasting effect on economic thinking, there has been a more recent upsurge in critiques of *homo economicus*. Economists George Akerlof and Rachel Kranton (2000), borrowing from the literatures in other disciplines, introduced into the economics literature the idea that some actions may be motivated by the desire to preserve one's identity—one's sense of who one is in the world—even when this is economically costly. Samuel Bowles, Ernst Fehr, and a number of other economists have, using mostly experimental techniques, investigated how regard for others and for the social meaning of an interaction affect people's choices (Fehr and Falk, 2002; Henrich et al., 2004). Richard Thaler was awarded the 2017 Nobel Memorial Prize in Economic Sciences in part for his recognition of the importance of people's concerns about fairness (Appelbaum, 2017). Economists seem to be belatedly "discovering" the social nature of human beings.

Highly relevant to the discussion of Figueres's approach is another important finding in the literature: there is growing evidence that *assuming* self-interest, teaching about *homo economicus*, and designing mechanisms to harness self-interest may actually serve to *encourage* selfishness. Some have found that the study of economics in the classroom discourages cooperation (Frank et al., 1993). Financial incentives have been found to, in some cases, "crowd out" intrinsic motivations (Frey, 1997). A number of behavioral economics studies have indicated that creating environments that signal that selfishness is appropriate can actually reduce pro-social choices (Bowles, 2008).

Such evidence suggests that Figueres's statements—which clearly signal that self-interest in the realm of international negotiations is appropriate and legitimate—may be seriously counterproductive.

Why did economists choose such a narrow model of human behavior? Notice that the image of "economic man" prioritizes autonomy, reason, mind, and separation from others and from the rest of nature. Notice also that the discipline's methods express a preference for precision and mathematical elegance, and that the discipline has defined itself around markets and/or choice-making.[1] Now notice what aspects of (all) human experience have been left out: connection and care, emotion, body, interdependence with others, interdependence with "mother" nature, richness, qualitative methods, home and community, and need. Lastly, notice the gender connotations of these two lists.

Feminist economists have been pointing out the partial and strongly gender-biased nature of mainstream economics since the 1990s (Ferber and Nelson, 1993; Nelson, 1997). It is not putting it too strongly to say that the professions obsession with self-interest and choice, to the exclusion of attention to the environmental embeddedness of the economy and to concerns of care for the environment and future generations, has strong roots in cultural sexism. Creating an adequate response to the climate crisis will require rehabilitating notions of ethical demands (as opposed to preferences) and powerful care (applicable across all spheres of human activity) that overcome sexist habits of thinking (Nelson, 2013; Nelson, 2016).

6.5. Climate Justice: What Really Motivates?

What really motivates? Let us look at some examples, both personal and national.

In 2017, US senior diplomat David H. Rank decided to resign rather than present the Trump administration's decision to withdraw from the Paris Agreement to the leaders of China, the country in which he was serving. He explained his reasons in an interview reported by the *New Yorker*:

> I'm not a great theologian, but, just in my gut, I thought, We're stewards of creation and the world. As a parent, I've spent my life trying to make my children's lives O.K. And, finally, in terms of national interests, it's just dumb.
>
> (Osnos, 2017)

[1] The issues of model and method are not unrelated, as the *homo economics* image was first suggested by John Stuart Mill as a way of making economics more "scientific," by which he meant more like geometry. To Mill's credit, he also noted that no one should be so foolish as to try to apply this model without recourse to the findings of other disciplines and to practical knowledge (Mill, 1836).

Notice the order of the reasons he gives. First comes sanctity, related to his sense of his relation to the greater universe. Second comes responsibility, related to his deeply felt identity as a parent. National interests and rationality are mentioned, but given last place. His personal economic self-interest was not served at all, since he felt so strongly about the wrongness of the administration's decision that he gave up a career of more than twenty-five years.

As another example, consider the writings of Hans Jonas in his book *The Imperative of Responsibility: In Search of Ethics for the Technological Age* (1984). While he spends considerable time discussing rational principles, Kant, Bacon, Marx, and various philosophical stances, he in the end claims a very different source for the type of ethics we need to address problems including environmental degradation:

> [A]ll proofs of validity for moral prescriptions are ultimately reduced to obtaining evidence of an "ontological" ought... [W]hen asked for a single instance... where the coincidence of "is" and "ought" occurs, we can point at the most familiar sight: the newborn, whose mere breathing uncontradictably addresses an ought to the world around, namely, to take care of him [*sic*]. Look and you know.
>
> (Jonas, 1984, pp. 130–1)

Jonas writes that "the always acute, unequivocal, and choiceless responsibility which the newborn claims for himself" creates "the ought-to-do of the subject who, in virtue of his power, is called to its care" (Jonas, 1984, pp. 134, 93, masculine pronouns in original). This claim points, not to the self-interest of the autonomous adult, but to a recognition of the infant's overwhelming aliveness and extreme fragility. It is grounded in a visceral perception of the deep interdependence of life, and the totally inegalitarian distribution—between the parent and the newborn—of the power to act to support and sustain that life.

While economic self-interest obviously often plays an important role in national actions, so do other motivations—for good as well as for ill. While, for example, a war may have roots in battles for the control of petroleum resources, and be fed by the interests of powerful weapons industries, few would willingly pay taxes to support the effort, much less volunteer to risk death on the battlefield, under the flag of ExxonMobil or Lockheed Martin. Moral motivators such as the Defense of Freedom, National Honor, Patriotism, Aiding our Allies, Defending our Homeland, Helping the Weak—as well as simply caring for one's buddies in the trenches—are always part of the story. Looking at how nations act, or how they "spin" or explain their actions, it is clear that nations seek respect, love, honor, and gratitude on the world stage. If they cannot get recognition through economic power or, say, hosting an Olympics event, then sucking their economy dry in order to wage a war or develop a nuclear weapon may seem like a viable option. Nations also often act, or present their actions, in such a way as to avoid guilt, shame, blame, embarrassment, disrespect, and hatred.

In 1977, US President Jimmy Carter gave a speech in which—drawing on the ideas of early twentieth-century philosopher William James—he likened the 1970s energy crisis to the "the moral equivalent of war" (Reston, 1977). While the presenting problem has changed since the 1970s, the question remains the same: how can we inspire, for peaceful ends, the same intensity of commitment all too often turned toward violent ones? Likening climate change decisions to cool financial calculations is not only unlikely to inspire such commitment, it may actively discourage it. Changing the rhetoric to focus on alternatives deserving of respect and honor provides a more promising route. Identifying caring as the strong and honorable thing to do needs to be the project of those concerned with climate justice.

6.6. Conclusion

Figueres's words opened this chapter, and will also end it. Figueres visited my university campus in April 2016, participating in two events. In the first, a small group discussion, she repeated her widely publicized, economistic claims about the importance of national self-interest. But an hour later she gave a thoroughly impassioned, and deeply compassionate, public lecture. Her theoretical (and, I have argued, unfounded) claims about "humans" and "nations" being primarily motivated by economic self-interest contrast quite markedly with how she explains her *own* motivations. An interview with *The Guardian* recounts the reasons she shared in her talk:

> Asked why she chose to work on climate change, she tells a story about the once common golden toad that went extinct in Costa Rica in 1989. Figueres has illustrations of the toad on the wall of her office.
>
> "I saw this species when I was a little girl, but when I had my two little girls the species no longer existed," she said. "It had just a huge impact because I realised that I was turning over to my daughters—who were very, very young, they were born in 1988 and 1989—a planet that had been diminished, by our carelessness, by our recklessness." It was this realisation that led her to work on the climate problem...
>
> Normally quick to respond to questions, the only time in our conversations that she hesitates is when asked about the fate of poor countries if the Paris talks fail. "I hope we don't fail," she eventually says. "They will be the ones that suffer."
>
> She looks away. There are tears in her eyes. (Harvey, 2015)

References

Adolphs, Ralph. 2009. "The Social Brain: Neural Basis of Social Knowledge." *Annual Review of Psychology* 60: 693–716.

Akerlof, George A., and Rachel E. Kranton. 2000. "Economics and Identity." *Quarterly Journal of Economics* 115 (3): 715–53.

Appelbaum, Binyamin. 2017. "Nobel in Economics is Awarded to Richard Thaler." *New York Times*, 9 October.

Bowles, Samuel. 2008. "Policies Designed for Self-Interested Citizens May Undermine 'The Moral Sentiments': Evidence from Economic Experiments." *Science* 320: 1605–9.

Bowles, Samuel. 2016. *The Moral Economy: Why Good Incentives Are No Substitute for Good Citizens*. New Haven, CT: Yale University Press.

CBS News. 2015. Video of interview with Christiana Figueres, October 2.

Fehr, Ernst, and Armin Falk. 2002. "Psychological Foundations of Incentives." *European Economic Review* 46 (4–5): 687–724.

Ferber, Marianne A., and Julie A. Nelson (eds.). 1993. *Beyond Economic Man: Feminist Theory and Economics*. Chicago: University of Chicago Press.

Fiske, Susan T. 2010. *Social Beings: A Core Motives Approach to Social Psychology*. New York: John Wiley & Sons.

Frank, Robert H., Thomas Gilovich, et al. 1993. "Does Studying Economics Inhibit Cooperation?" *Journal of Economic Perspectives* 72: 159–71.

Frey, Bruno S. 1997. *Not Just for the Money: An Economic Theory of Personal Motivation*. Cheltenham: Edward Elgar.

Greene, Joshua, and Jonathan Haidt. 2002. "How (And Where) Does Moral Judgment Work?" *Trends in Cognitive Sciences* 6 (12): 517–23.

Harvey, Fiona. 2015. "Christiana Figueres: The Woman Tasked with Saving the World from Global Warming." *The Guardian*, 27 November.

Henrich, Joseph, Robert Boyd, et al. (eds.). 2004. *Foundations of Human Sociality: Economic Experiments and Ethnographic Evidence from Fifteen Small-Scale Societies*. Oxford: Oxford University Press.

Herzberg, Frederick. 1987. "One More Time: How Do You Motivate Employees?" *Harvard Business Review* 65 (5): 109–20.

Jonas, Hans. 1984. *The Imperative of Responsibility: In Search of Ethics for the Technological Age*. Chicago: University of Chicago Press.

Mill, John Stuart. 1836. "On the Definition of Political Economy; and On the Method of Philosophical Investigation in That Science." *London and Westminster Review* 426: 1–29.

Nelson, Julie A. 1997. "Feminism, Ecology and the Philosophy of Economics." *Ecological Economics* 20: 155–62.

Nelson, Julie A. 2013. "Ethics and the Economist: What Climate Change Demands of Us." *Ecological Economics* 85 (January): 145–54.

Nelson, Julie A. 2016. "Husbandry: a (Feminist) Reclamation of Masculine Responsibility for Care." *Cambridge Journal of Economics* 40 (1): 1–15.

Osnos, Evan. 2017. "Wind on Capitol Hill: Homecoming." *New Yorker*, 3 July.

Posner, Eric A., and David Weisbach. 2010. *Climate Change Justice*. Princeton, NJ: Princeton University Press.

Reston, James. 1977. "Moral Equivalent of War." *New York Times*, 20 April: A25.

Sen, Amartya. 1977. "Rational Fools: A Critique of the Behavioral Foundations of Economic Theory." *Philosophy and Public Affairs* 6 (Summer): 317–44.

Tabuchi, Hiroko. 2016. "U.S. Companies to Trump: Don't Abandon Global Climate Deal." *New York Times*, 16 November.

7

Noncompliers' Duties

Anja Karnein

When a country, such as the United States, leaves a symbolically important international climate change mitigation agreement in a celebratory and self-congratulatory fashion, it is failing to comply in at least two respects. Most obviously, it is failing to fulfill its duty to honor the agreement, which is arguably part of its climate change mitigation obligations.[1] Less obviously, but still importantly, it is failing to behave decently as a noncomplier, by further undermining public morale on climate change mitigation. There are various ways to explain what makes it wrong for a noncomplier to behave in this way. One might suggest that there are further general obligations, such as the duties to behave decently or to minimize the harm one is causing, which the noncomplier, just as anyone else, should respect. In this chapter I offer an alternative explanation. I propose that a category of duties one might call "noncompliers' duties" is better able to capture the intuition that a behavior such as that of the United States is particularly problematic, more so, that is, than it would be if it entailed simply violating two separate duties. Noncompliers' duties are moral requirements that arise with special force for the noncomplier *because* they failed to comply.

This may strike the reader as a strange way of thinking about noncompliance. The more familiar way is to consider how the duties of compliers change in the presence of noncompliers (Murphy, 2000; Horton, 2004, 2011; Hohl and Roser, 2011; Miller, 2011; Karnein, 2014; Stemplowska, 2016). In contrast, the question what duties noncompliers may have in light of their failure to comply has thus far not received much attention. This both is and is not

[1] Even though, as everyone knows, the duties specified in the agreement are the outcome of political bargaining and as such barely map on to what is morally required of each country in terms of climate change mitigation.

surprising. It is surprising in the sense just described, namely that it seems intuitively worse if a noncomplier behaves badly than if a complier does.[2] At the same time, the failure in the literature to acknowledge noncompliers' duties is not surprising since, as this chapter demonstrates, establishing such duties is more complicated than one might think.

The argument proceeds in two steps. I begin by briefly setting the stage and explaining the problem with establishing noncompliers' duties, especially in situations where compliance is still an option. I then go on to explore four ways of capturing common intuitions about duties noncompliers have. The first accounts for duties we might think noncompliers have by postulating a general duty to produce the most good given how much moral agents are prepared to do, a duty which may be reiterated whenever noncompliance occurs. The second proposes that, once an agent chooses not to comply, the morally relevant circumstances change so that an entirely new duty emerges for that agent. The third suggests that all duties that may seem to "arise" for the noncomplier are actually always already present. The fourth explains duties noncompliers have as manifestations of a "leftover part" of the duties that were not complied with, namely the imperfect duty of beneficence, which changes from discretionary to non-discretionary in the context of noncompliance. I show that all four explanations are riddled with difficulties. Still, I claim that the fourth option most plausibly captures the intuition that noncompliers ought to behave better rather than worse *because* they are noncompliers. Before I begin, please note that this chapter should be read as exploratory. It will make some suggestions in order to start a discussion, none of which are meant as the final word on the subject.

7.1. The Status Problem

Noncompliance can occur either in situations in which compliance is still an option or in situations in which it is not.[3] The most familiar and uncontroversial set of noncompliers' duties arises in the latter situation, that is, when the initial duty-triggering situation has passed. Here we expect noncompliers to take responsibility for their wrongful (in)actions and thus regret, and

[2] Notable here is also the debate among deontic logicians about contrary-to-duty imperatives, where it is taken to be a problem of deontic logic that intuitively plausible noncompliers' duties cannot be established, or at least not without contradiction: Chisholm, 1963a; Aquist, 1967; Sellars, 1967; Forrester, 1984; Saint Croix and Thomason, 2014.

[3] Noncompliance may no longer be an option because the initial duty-triggering problem has resolved itself: either because someone else has taken up all of the slack or because nobody has done anything and the state of affairs to be prevented has already materialized.

apologize as well as perhaps compensate for, their failure. Criminals may even be thought to have a duty to accept punishment (Tadros, 2011). These duties arise exclusively for the noncomplier in direct response to not having done what she ought.

The more problematic case ensues when the initial duty-triggering situation is ongoing and compliance remains an option,[4] as it is with current duties concerning climate change. In such situations there are two different sorts of behaviors we might think noncompliers ought to display. First, where possible, they ought to fail to comply as little as possible. Depending on the case, this may mean changing their course of action entirely (Lipsey and Lancaster, 1956) or doing as much of the original action as they are prepared to do. We might think, for instance, that a country always has more reason to do 90 percent of what it should in terms of climate change mitigation efforts than, say, 60 percent. I call this the *approximation requirement* (*appr*).

Second, of the various behaviors agents can choose from when failing to comply, they ought to choose the most decent behavior.[5] This category contains duties that do not allow partial compliance in the way spending on emissions does. That is, a duty such as "do not kill" is absolute. There is no way of killing a person by only 90 percent or 60 percent. In such cases, where the noncomplier is automatically committed to failing entirely, or complying with 0 percent, we might think the noncomplier ought to behave better rather than worse at her level of compliance (which may amount to 0 percent). The murderer could, for instance, act so as to be:[6]

b_{nc1}:[7] "gentle," by making an extra effort to kill quickly and painlessly,
b_{nc2}: "neutral," by killing without making this particularly quick or painless but also without trying to prolong it unnecessarily or inflicting extra pain,
b_{nc3}: "sadistic," by killing carelessly, thereby taking more time and inflicting extra pain.

[4] This category is usually only implicitly—if at all—distinguished from that where the initial duty-triggering situation has passed (Chisholm, 1963a, p. 36).

[5] Dominic Roser has proposed that partial compliers have the duty to maximize the efficiency of their actions within the limits of their only partially compliant motivation (Roser, 2016). In other words, a country that is committed to only spending 60 percent of what would be required to reduce emissions, for instance, should be spending those 60 percent in a more rather than less efficient manner. This allows the same amount of expenditure to produce more good. The problem, however, is that it remains unclear what the appropriate criterion is for what compliance demands. Are we asking agents to spend a certain amount of money (or effort) or to produce a certain amount of good?

[6] The following is an expanded version of Forrester's gentle murder example (Forrester, 1984).

[7] "b_{nc}" is short for "possible behavior of noncomplier."

Another frequently cited case is that of the broken promise. Suppose you have promised to visit your friend on Sunday.[8] If you break your promise you have at least three behaviors to choose from. You could, for instance, be:

b_{nc1}: "caring," by calling ahead and announcing that you will not be coming,
b_{nc2}: "neutral," by just not showing up,
b_{nc3}: "callous," by just not showing up after having already having made the call saying when to expect you.

We might think that it is the duty of the noncomplier to choose the "gentle" and "caring" options, respectively. I call this the *amelioration requirement* (*amel*). With regard to both categories, while it seems obvious *that* noncompliers should choose certain options over others, it is less obvious *why* this should be so. Perhaps the original duty reads something like:

d_o: Do X.
d_{nc}: If you fail to do X,
(*appr*) limit your failure to do X,
and, within your level of noncompliance,
(*amel*) behave decently.

The problem with this is that there seems to be something seriously wrong with a duty that includes prescriptions for what to do once it is not fulfilled: for what exactly is your duty? Is it to fulfill your duty or is it to fail decently? The problem is the status of any potential noncompliers' duty with respect to the original duty in cases in which compliance is still an option. I refer to this as the *status problem*. This has to be addressed by any view attempting to establish duties for noncompliers in situations where the initial duty-triggering situation is ongoing and compliance remains a possibility.

7.2. Four Ways to Think of Duties Noncompliers Have

In the following I examine four possible ways to explain the duties we might think arise for agents who fail to comply in cases where compliance remains an option. The first three avoid the status problem but, they do so largely, at the cost of not being able to establish genuine noncompliers' duties, that is, duties that arise for the noncomplier *because* she failed to comply. To the extent any of these first three views are convincing, we might conclude that while noncompliers may remain duty-bound this is not due to their noncompliance and, thus, that the notion of noncompliers' duties as a separate category of

[8] This is a variation of an example used by Chisholm (1963) and Sellars (1967).

duties is chimerical. The fourth position, however, does at least come close to establishing duties that arise *because* of noncompliance. This, or so I maintain, is also the most plausible account. To the extent, therefore, that this position is sound and can indeed best capture our intuitions, there may be a useful category we could refer to as noncompliers' duties.[9]

7.2.1. *The Assimilation View*

The *assimilation view* is a nonideal consequentialist view and formulates the original duty so that it can capture most cases of what we might think are noncompliers' duties. Its basic commitment is to the following ideal duty:

d_o: Always act so as to produce the most good.

To this it adds the assumption that always producing the most good entails not asking of agents to always produce the most good possible, but that producing the most good requires working within the given motivational set-up of agents (Roser, 2016; Hooker, 2000). Roser, for instance, could be read as establishing the following original, but nonideal, duty:

d_{o*}: Always act so as to produce the most good given the level of effort and sacrifice you are prepared to make.

This would clearly explain why we might think that noncompliers have duties: even if they are not prepared to do all that might be required to produce the most good (d_o), they remain under an obligation to produce the most good given what they are willing to do (d_{o*}). It can thereby account for the amelioration requirement, namely why we, for instance, would want the murderer to kill gently and the promise-breaker to act caringly (assuming that both actions do, indeed, produce the most good).

By contrast, it cannot explain the approximation requirement, that is, why it would be better to cut emissions more rather than less, irrespective of the amount of sacrifice an agent is prepared to make. That is, if the agent is only willing to do 60 percent there is no point—it may, in fact, be deemed to be counterproductive—to ask for more. Moreover, by assimilating d_o and d_{o*}, there really is only very little space for noncompliance. Not complying would mean failing to produce even the amount of good one was prepared to produce. On this account most agents will thus end up being compliers of

[9] I should note that I am more or less making up the following four positions as a way of trying to systematize various suggestions within the literature to explain duties noncompliers may have. These positions do not claim to be exhaustive but simply a first attempt at trying out several possibilities.

sorts. Anyone who fails to comply with even this personally tailored duty would need a separate prescription for what to do:

d_{o^*}: Always act so as to produce the most good given the level of effort and sacrifice you are prepared to make.
d_{nc}: If you fail to do that, act so as to produce the most good you are prepared to produce from then on.

This, however, would immediately raise the status question again, just on a lower level. The explanation for noncompliers' duties must therefore be different than what the assimilation view, as I am portraying it, has to offer. The next approach does not attempt to radically reduce the possibilities for noncompliance, although it does seek to make noncompliance a one-off event that leads to new original duties.

7.2.2. *The Supersession View*

The *supersession view* starts by asking the agent to always fulfill her original duty. She can, of course, fail to do so. But if she fails, her noncompliance creates new facts on the ground. She is now presented with new circumstances that require her to ask afresh: what ought I do? She finds herself with a new original duty that supersedes the previous one. Depending on which sort of behavior is at issue, this duty is likely to be either quantitatively or qualitatively the next best thing to the original duty (e.g. 99.99 percent) or it will respond to the new situation created by the noncompliance (murder gently since murder is what you are going to do). It could be formulated thus:

d_o: Do X.
If you fail to do X,
ask afresh: what ought I do given the new facts on the ground (i.e. my failure to do X)? If the answer is Y then
d_o: Do Y.

This view addresses the status problem by not ever allowing two different duties to "hover" over the agent: whenever she fails to comply, that changes the circumstances so that duties become redistributed. This reduces the possible behaviors of noncompliers to one in each instance and opens up a multitude of original duties, each of which newly arises whenever an agent fails to comply. Note that this solves the problem by treating the case in which the initial duty-triggering circumstance is ongoing just like the case in which the initial duty-triggering circumstance has passed (see Section 7.1).

While this view can account for the expectation that noncompliers remain duty-bound, it is not a particularly plausible position. It relies on "pretending" that compliance is no longer an option when it actually is. It resolves the conflict by artificially removing the option to comply from the noncompliers' catalogue of choices and just asking the agent to re-pose the question "what ought I do?" whenever she has chosen not to do what she ought to do.

There is also a problem with referring to the newly acquired duties as original duties, at least if this is taken to indicate that previous noncompliance plays no role in the assignment of the duty—which it would have to do, since, again, for this view it is necessary to treat all situations past the moment of noncompliance as "new," that is, unburdened by previous behavior. The problem is that "murder gently" obviously cannot be a duty for anyone but a murderer (who is clearly a noncomplier).

Thus, insisting that what may appear to be noncompliers' duties are actually a myriad of serially superseding original duties does not persuasively capture what we intuitively consider to be duties noncompliers have. The discussion of the supersession view, which assumes that new duties spring up constantly, has, however, provided some clues for where to turn to next, namely to explore the possibility of the opposite view, which posits that all duties one may ever have are always already present.

7.2.3. *The Ubiquitous Duties View*

The *ubiquitous duties* view explains what we may think are duties noncompliers have by showing that these are duties everyone always already has. That is, an agent has both the (d_o) duties not to murder and to be gentle. Thus, if an agent violates the first of these two duties all that happens is that the second remains. It could be formulated thus:

d_{o1}: Do X.
d_{o2}: Do Y.
d_{o3}: Do Z.
If you don't do X, you still have to do Y and Z.

This view addresses the status problem by denying that there is ever a conflict between the various duties: they are all always present and if an agent fails to fulfill one, her duty to do so remains, as does her duty to do all the other things she may have to do in a given situation.

While this solution may initially look plausible, it cannot adequately explain the approximation requirement: doing what was previously 99.99 percent of a duty can only ever be a duty for the noncomplying agent and arise in the context of a previous failure to comply. That is, if we think that an agent refusing to lower emissions by 100 percent has reason to lower it by 99.99 percent then we would have to think, on this view, that there are innumerable incremental duties between 0 and 100 percent. In other words, one would have to maintain that an agent not only has the duty to lower emission by 100 percent but also by 99.99 percent and by 99.98 percent and by 99.97 percent, etc. That seems implausible.

It also runs into difficulties when it comes to the amelioration requirement. It seems unable to explain how "be gentle" or "be caring" can be general duties that are always already present. While apparently quite general, they are still different from separate original duties such as "you ought not steal" or "you ought not lie." The latter obviously apply to everyone, compliers and noncompliers alike. If you kill, you still ought not steal or lie. If you steal, you ought not kill or lie, etc. But compliers do not always have to be gentle. While gentleness may be a generally favorable mode of interacting with others, sometimes—and unlike the duties not to steal and not to lie—gentleness may not be owed, for instance, when it comes to interactions with noncompliers.

The reader might point out that this should not come as a surprise: duties to be gentle or caring are imperfect duties, while duties not to steal or lie are perfect. As imperfect duties, while they are as general as the perfect duties, there is some discretion on the part of agents when to act on them.[10] So suppose agents always have the imperfect duty to be gentle and caring so that, when they fail to comply with some duty, they still have reason to at least consider being gentle or caring. While this implies that all agents always have the imperfect duty to consider being gentle and caring, it also suggests that they can always decide not to act on these duties in any given circumstance. This could indeed explain why a noncomplier acted gently or caringly after the fact ("she acted on her imperfect duty to be gentle or caring"). But it cannot explain why, in any given circumstance, the noncomplier *should* act this way. However, as I will show in Section 7.2.4, the route via imperfect duties is promising—it just does not work to explain noncompliers' duties if we think of them as general duties all agents always already have. Before I turn to spelling this out in more detail, let me consider one more way to salvage the ubiquitous duties account.

One might claim, for instance, that both the approximation and the amelioration requirements can be explained by a general—separate, always already present—duty to reduce the harm one is causing. This would apply to all

[10] Imperfect duties allow a certain degree of latitude. The agent may choose when and where to act on them, as long as she acts on them sometimes and to some extent (Hill, 1971, p. 56). Here the crucial question is on what basis the agent can decide—whether we take this to be more or less by whim (as long as she acts on it sometimes and to some extent, see Hill, 1971) or by some more stringent criteria that makes the decision less a matter of mere inclination and more a matter of judgment (Herman 1984). Although I have no space to argue for this here, the latitude seems to be more credibly reconstructed as a matter of judgment about whether one ought to act on one duty rather than on another as opposed to as allowing for inclination to trump acting on duty.

agents, compliers and noncompliers alike, since all agents can, on occasion, cause harm. While this may appear like a viable solution, it cannot explain why we might think that certain ways of causing harm, such as via noncompliance, are worse than others and warrant different responses: consider a variation on the case of the promise-breaker. Suppose you have promised a friend to join her at a concert on Sunday. She has cancelled all other plans in order to be able to go with you. If you choose not to join her now (although you still could), you arguably have a strong duty to call her and tell her in advance. Even if you do not have her number or other way to contact her, it is your duty to exert some effort to find her information. Now compare this to the case in which your friend, due to no fault of yours, comes to believe that you will be at the concert on Sunday (perhaps your friend likes reading tea leaves). It is Sunday morning, you know that for reasons you had nothing to do with your friend expects you to be at the concert. But you do not want to go. There seems to be no reason for you to call her and tell her this, especially not if you would have to exert effort to find her contact information. According to the ubiquitous duties view, however, complier and noncomplier alike have a duty to reduce the harm they are causing and, in both cases, we are assuming that your absence at the concert on Sunday is causing harm. But in this case it would not seem to be true that there is a general, always present, duty to reduce the harm one is causing.

Thus, a view holding that all duties agents ever acquire are always already present runs into two sorts of difficulties. First, if it posits always already existing imperfect duties, such as to be gentle or caring, then it cannot explain why noncompliers should choose, in any given instance, to act on them. Second, even if there was a general duty to minimize the harm one is causing, the view cannot distinguish the different sorts of responsibilities (and thus, duties) we may intuitively think noncompliers have but that compliers do not. The final view seeks to address these shortcomings by proposing not that all duties one ever has are always already present for all moral agents, but that all duties agents have are accompanied by an implicit, imperfect duty of beneficence.

7.2.4. *The Complementarity View*

The *complementarity view* accounts for the duties we think noncompliers have and addresses the status problem by proposing that all duties agents have are always complemented by an implicit, imperfect, duty of beneficence, that is, to make the ends of others one's own. This suggestion reinterprets the duty of beneficence as not just having its own sphere of duties separate from (perfect) duties of right, as Kantians might suggest, but as also implicitly accompanying or, more precisely, specifying, all duties, be they perfect or imperfect. In outline, the idea, which I spell out in more detail below, is this:

> Every duty consists of two parts:
> d_{o1}: whatever the duty explicitly entails (i.e. do X) and
> implicit d_{o2}: whatever mode of fulfilling the duty affirms your commitment to making the ends of others your own
> If you fail to fulfill d_1, d_2 remains. Thus, you ought to choose whatever mode of not fulfilling the duty affirms your commitment to making the ends of others your own, which will include:
> (*appr*): limiting your noncompliance,
> and, within your level of noncompliance,
> (*amel*): behaving beneficently.

In contrast to the ubiquitous duties account, on this view I do not have two separate duties not to break my promises and to be beneficent, which makes it the case that once I break my promise to visit my friend I still have to consider being beneficent but could choose, on any given occasion, not to be. Rather, I have the duty to choose, among the various options of keeping my promise, the one that is beneficent. In other words, I ought to call ahead to announce my visit and not just show up. I could, of course, always choose not to act on the beneficent part of my duty in any specific instance. If I break my promise, however, the duty of beneficence remains. And in the context of my currently noncompliant behavior, or so I propose below, I am no longer as free as I was before to choose a nonbeneficent mode of conduct.

This claim makes a number of controversial assumptions. For that reason alone the ubiquitous duties view might be preferable, even at the cost of not being able to fully capture our intuitions. However, let me try to develop the argument for the complementary view step by step to see if I can rescue it from obscurity. In the following I first explain (a) what it means to say that a duty consists of two parts, and second (b) why or how one part remains after the other has not been complied with. The final step consists in demonstrating (c) that what I propose are indeed noncompliers' duties in the sense of arising for the noncomplier *because* she failed to comply.

(a) The basic idea of ascribing to duties an implicit part is this: although it is imperfect duties that are known for the latitude they allow, most duties, even perfect ones, also allow a certain degree of latitude in how they are fulfilled. One example for this that has been used in the literature is the duty to pay your debt (Chisholm, 1963b, p. 4; Hill, 1971, p. 63). This will not necessarily specify how you are to do this, by cash, check, or money order. Which of these the agent chooses may sometimes—as it presumably is in this case—be morally irrelevant. But it would be surprising if it always were. Thus, no matter how small or insignificant this latitude is thought to be, as long as there is some such latitude, and as long as one is not willing to say that it is always morally immaterial how the duty is to be fulfilled, there also needs to be a principle to specify what the agent has most reason to do. I am suggesting that it is the duty to make the ends of others one's own, implicitly accompanying all duties, which does this.

Mavis Biss suggests something similar (Biss, 2017). She claims that, counter to a prominent view in the literature, it is not a merely prudential decision how to fulfill one's perfect duty not to lie, for instance. The agent can presumably choose among a wide variety of ways to fulfill the duty. She can, for example, say nothing or give highly selective answers to certain questions. Or she could be truthful. Biss claims that because the duty not to lie is informed by its positive counterpart, namely truthfulness, the discretion an agent may be thought to have with regard to deciding between the various ways of not lying are limited by what my duty to be truthful will allow.[11]

My idea about a duty of beneficence complementing all duties is similar, to the extent at least that it also suggests that there is more to fulfilling duties than merely not to violate them.[12] That is, there may be morally permissible and morally impermissible ways of fulfilling them, as strange as this may initially sound. Let me illustrate this with an example for how such an implicit duty may specify a duty that *is* being complied with. Consider the case of curbing emissions and, fantastically, Fulfillistan, a fully compliant country. It could comply in various ways, acting so as to be:

b_{c1}:[13] "helpful," by, for instance, announcing publicly and proudly how easy it is to comply, thereby boosting global morale on climate change mitigation,
b_{c2}: "neutral," by simply cutting emissions without making any public statements,
b_{c3}: "obstructive," by being compliant but letting the public think that it is just complying because it is being strong-armed, thereby undermining global morale on climate change mitigation.

To the extent it is plausible that b_{c1}–b_{c3} are all within the scope of the original duty to cut emissions (I discuss this assumption below), I am proposing that it is the duty of beneficence, which specifies which of the three one ought to do—in this case one should clearly choose b_{c1}.

[11] Within what is truthful, there may, of course, again be some latitude, but the choice here can then be made on prudential grounds.

[12] In most other respects my proposal is quite different. Most notably, perhaps, it is not an attempt to interpret Kant in the most plausible way. I also suggest that there is one duty, namely beneficence, that specifies duties and not that it is the positive counterparts of duties that do this, as Biss claims.

[13] "b_c" is short for "possible behavior of complier."

(b) When an agent turns into a noncomplier by having failed to curb her emissions to the appropriate extent, this duty remains, at least in its basic appeal to the agent to make the ends of others her own. It remains because arguably that part of the duty can still be complied with—an agent can still make the ends of others her own, even with regard to the behavior that comes with the decision not to comply with a particular duty. Sometimes this appeal will translate into retaining the duty exactly. Consider, for instance, a country that lowers its emissions by only 60 percent of what it ought to. Here the duty to publicly embrace the global effort to curb emissions will remain as it is. At other times, the basic appeal will translate into altered behavior. Consider again the promised visit to your friend. I am claiming that beneficence gives you a reason to call ahead and announce your visit as opposed to just showing up. Now imagine that you break your promise: you will not visit your friend sometime on Sunday. It then makes no sense for you to make this particular courtesy call (telling her you are on your way) that beneficence would have required of you as a complier. But if the duty of beneficence remains, which it arguably does since you are "only" breaking the promise to visit on Sunday—and, I am assuming, do not generally intend to undermine your friendship—then beneficence will require you to make a different courtesy call, namely to announce that you will not be coming.

The reader might see various problems with this suggestion. One objection I have already hinted at might be that b_{c3} cannot plausibly be understood as within the scope of the perfect duty to curb emissions. That is, the duty itself already contains a certain level of care that would exclude being obstructive in the way described. But this is not at all clear: the duty to "curb emissions" may amount to just that, the duty to curb emissions. The objection is certainly right in pointing out that this will not condone any means possible. But it will be hard to make the case that cutting emissions will be morally acceptable only if it is done with the display of the right public attitude.

One might also think the opposite, namely that displaying the right public attitude is a separate $duty_o$ (like not to kill, not to steal, and not to lie, see the discussion in Section 7.2.3). But even if it is (and this example was poorly chosen), and even if displaying the right public attitude turns out to be a separate original duty, it will be difficult to deny that some latitude for fulfilling duties will often remain and that this latitude will contain more or less desirable actions, even if the degree of difference (between helpful and obstructive) can be quite minimal.

For argument's sake, therefore, assume that none of the behaviors described in b_{c1}–b_{c3} are such as to threaten the status of the action as fulfilling the duty. Then it seems that the only way to insure that a complier chooses b_{c1} over the other possibilities is to complement her duty with one demanding a certain degree of beneficence in carrying out the duty. Engaging the duty of beneficence this way would give agents a reason to prefer some ways of fulfilling a duty over others, depending on the extent they entail making the ends of others their own above and beyond what the merely formal way of fulfilling the explicit duty may or may not do.

At this point a critic might see further problems with this, especially when it comes to understanding what beneficence means for noncompliers. She might, for instance, note that once someone fails to fulfill her duty to choose b_{nc1}, for instance, she will be a noncomplier on both scores (with regard to both d_o and d_{nc}) and thus will, again, not have reason to choose the more beneficent of the remaining options (b_{nc2} over b_{nc3}). Likewise, it is unclear what beneficence would have to say to someone who, once she has failed to do 100 percent of her duty, such as cut emissions, also fails to do the next best thing, such as cut emissions by 99.99 percent.

But arguably, beneficence comes in degrees: while one certainly has more reason to display more of it than less, not doing more does not preclude doing only a bit (as opposed to a lot) less. In general, the agent must make sure that across her choices she displays "enough" beneficence so that her general commitment to this end does not come into question. If she persistently chose to do barely more than not violate the explicit parts of her duties, it would be hard for her to make the case that she is generally committed to making the ends of others her own.

However, and now comes a crucial step for my argument, I want to maintain that the persistent failure of a complier to act on her (implicit and complementary) duty of beneficence does qualitatively less to undermine her credibility as someone who generally is committed to the duty of beneficence than does the failure of a noncomplier to act on this duty. The problem for the noncomplier is that she has already done something that violates her duty, either to herself or to others. She has thereby violated the moral law, or, in non-Kantian terms, one might say her duties to the moral community. Choosing a beneficent noncompliers' option is thus a way for her to reassure all other moral agents that, while she is failing to comply in this instance, she remains generally committed to participating in the moral community. Choosing a nonbeneficent remaining option amounts to a serious aggravation of the first failing. She signals that "not only am I failing to comply in one particular instance, I am also making a more general statement that I really do not care for being part of the moral community." Thus, the less beneficent an agent chooses to be, the more disappointing and worrying her behavior becomes overall. The choice, therefore, to not be beneficent or even to land somewhere rather low on the beneficence range constitutes a separate, serious, step for the noncomplier.

If this is correct, then, once an agent has failed to fulfill her original duty, that is, has become a noncomplier, she remains subject to the duty of beneficence that guided d_o. This would give the noncomplying agent reason to prefer doing 90 percent of her duty over doing 60 percent and choosing b_{nc1} over b_{nc2}, b_{nc2} over b_{nc3}, and so forth—at least until a threshold is reached below which one can no longer speak of beneficence.

(c) But what, you might ask, makes these duties into noncompliers' duties? It is certainly not that duties of beneficence uniquely apply to noncompliers, which they obviously do not. Many—but, importantly, and different from what the ubiquitous duties view would have us believe, not all—actions ought to be done gently, for instance, not just murder. But the duty applies to noncompliers in a particular way, or, as I said earlier, in a qualitatively different way, given what a failure to fulfill it signals in the context of noncompliance: when a complier chooses a nonbeneficent option she remains a complier but sends a somewhat dismissive signal to those to whom the duty is owed. However, as also already mentioned, depending on her choices in other instances of compliance, this might do nothing to undermine her credibility of being generally committed to making the ends of others her own: beneficence, as an imperfect duty, does not ask her to necessarily always choose the most beneficent way of fulfilling a duty, especially not if other duties she has (toward her own self-perfection, for instance) happen to offer competing reasons.

This situation, or so I want to propose, is very different for the noncomplier. Here, similar perhaps to cases of obligatory aid, acting on the—still imperfect—duty of beneficence is not optional when choosing among the various possible modes of not complying (Herman, 1984; Stohr, 2011). A noncomplier has thus far only shown disrespect for those to whom the duty is owed and to the moral community more generally. She now has to provide some sort of reassurance that she wants to remain a part of the moral community. This is not unlike redemptive duties (to regret, apologize, compensate, etc.) that may be thought to work in cases where the initial duty-triggering situation has passed. Here nobody can undo what has been done, but symbolic gestures of remorse can, and ought to, be made to mend the damaged ties with the moral community. For noncompliers in ongoing duty-triggering situations it may be seen as a form of advance assurance of future allegiance despite not fulfilling the particular duty in question. Choosing a beneficent of the remaining options thus matters more in the case of the noncomplier than it matters in the case of the complier—so much more, that is, that the noncomplier largely loses the discretion to choose a nonbeneficent behavior.[14]

[14] The complier's discretion is not unlimited. Even on the most lenient reading, where the complier may choose, in any given instance, to forgo beneficence in favor of following her inclinations, there will come a point at which she will have to choose beneficence in order for her proclaimed commitment to the maxim of beneficence to remain credible.

Thus, it is this irreducible importance of beneficence in cases of noncompliance that makes it into a specifically noncompliers' duty. To return to the earlier example: it would be great if Fulfillistan publicly embraced climate change mitigation policies and thus boosted global morale. The imaginary country would have had reason to choose this, the most helpful, option. If it did not choose this but instead acted on either the neutral or obstructive alternative, Fulfillistan would simply provide no extra affirmation of those the duty is owed to and the moral community more generally. Beyond this, the nonbeneficent choice would not amount to much of a transgression—there may be other opportunities to affirm its commitment to the moral community. By contrast, it would be very concerning if a country that already fulfilled only 60 percent of its duty now also refused to be cooperative when it comes to boosting global morale or did something else that signaled its continued unwillingness to cooperate more generally.

7.3. Conclusion

I have argued that there is something particularly bad about an agent applauding herself for her failure to comply with her climate change mitigation-related duties. This badness, I maintained, cannot be fully captured by explaining the shortcoming as a violation of an independent duty, such as the duty to minimize the harm one is causing or the duty to behave decently. It is better captured by recognizing the violation of a noncompliers' duty—the duty for a noncomplier to choose, from the various options open to her, a beneficent one over others. This imperfect duty, I suggested, is a leftover part of the duty she is failing to comply with and, in the context of noncompliance, loses much of its optionality. It requires the agent to, for instance, publicly affirm the importance of climate change mitigation despite her own failure to live up to her duties in this regard. Fulfilling this duty signals to those to whom the duty is owed and the moral community as a whole that, while the agent is not doing what she ought, she is, in principle at least, willing to cooperate.

Acknowledgments

I am grateful to Mattias Iser, Christopher Morgan-Knapp, Matthew Rendall, Henry Shue, and Melissa Zinkin for their very helpful advice and suggestions.

References

Aquist, Lennart. 1967. "Good Samaritans, Contrary-to-Duty Imperatives, and Epistemic Obligations." *Noûs* 1 (4): 361–79.

Biss, Mavis. 2017. "Avoiding Vice and Pursuing Virtue: Kant on Perfect Duties and 'Prudential Latitude'." *Pacific Philosophical Quarterly* 98: 618–35.

Chisholm, Roderick. 1963a. "Contrary-to-Duty Imperatives and Deontic Logic." *Analysis* 24 (2): 33–6.

Chisholm, Roderick. 1963b. "Supererogation and Offense: A Conceptual Scheme for Ethics." *Ratio* 5 (1): 1–14.

Forrester, James William. 1984. "Gentle Murder, or the Adverbial Samaritan." *The Journal of Philosophy* 81 (4): 193–7.

Herman, Barbara. 1984. "Mutual Aid and Respect for Persons." *Ethics* 94 (4): 577–602.

Hill, Thomas. 1971. "Kant on Imperfect Duty and Supererogation." *Kant Studien* 62: 55–76.

Hohl, Sabine, and Dominic Roser. 2011. "Stepping in for the Polluters? Climate Justice under Partial Compliance." *Analyse und Kritik* 33 (2): 477–500.

Hooker, Brad. 2000. *Ideal Code, Real World: A Rule-Consequentialist Theory of Morality.* Oxford: Oxford University Press.

Horton, Keith. 2004. "International Aid: The Fair Shares Factor." *Social Theory and Practice* 30 (2): 161–74.

Horton, Keith. 2011. "Fairness and Fair Shares." *Utilitas* 23 (1): 88–93.

Karnein, Anja. 2014. "Putting Fairness in its Place: Why there is a Duty to Take Up the Slack." *The Journal of Philosophy* 111 (11): 593–607.

Lipsey, R. G., and Kelvin Lancaster. 1956. "The General Theory of Second Best." *The Review of Economic Studies* 24 (1): 11–32.

Miller, David. 2011."Taking Up the Slack? Responsibility and Justice in Situations of Partial Compliance." In *Responsibility and Distributive Justice,* edited by Carl Knight and Zofia Stemplowska, 230–45. New York: Oxford University Press.

Murphy, Liam. 2000. *Moral Demands in Nonideal Theory.* Oxford: Oxford University Press.

Roser, Dominic. 2016. "Reducing Injustice Within the Bounds of Motivation." In *Climate Justice in a Non-Ideal World,* edited by Clare Heyward and Dominic Roser, 83–103. Oxford: Oxford University Press.

Saint Croix, Catharine and Richmond Thomason. 2014. "Chisholm's Paradox and Conditional Oughts." In *Deontic Logic and Normative Systems,* edited by F. Cariani, D. Grossi, J. Meheus, and X. Parent, 192–207. Cham: Springer International Publishing.

Sellars, Wilfrid. 1967. "Reflections on Contrary-to-Duty Imperatives." *Noûs* 1 (4): 303–44.

Stemplowska, Zofia. 2016. "Doing More than One's Fair Share." *Critical Review of International Social and Political Philosophy* 19 (5): 591–608.

Stohr, Karen. 2011. "Kantian Beneficence and the Problem of Obligatory Aid." *Journal of Moral Philosophy* 8: 45–67.

Tadros, Victor. 2011. *The Ends of Harm: The Moral Foundations of Criminal Law.* Oxford: Oxford University Press.

8

Divest–Invest: A Moral Case for Fossil Fuel Divestment

Alex Lenferna

8.1. The Moral Complexity of Divestment

The fossil fuel divestment movement is inspired by a powerful history of students and communities calling for institutional investments to match the values of those institutions. Most prominently it is modeled after the anti-apartheid divestment movement, which demanded that institutions divest from companies operative in apartheid South Africa. While the demands of the fossil fuel divestment movement vary, the most common demand is for institutions to divest from owning shares in the 200 publicly traded fossil fuel companies with the highest amount of reported carbon reserves (AKA the Carbon Underground 200). There have also been divestment campaigns aimed at companies inconsistent with the Paris Climate Agreement, specific fossil fuel projects or sectors, such as tar sands and coal, bank financing of fossil fuel projects, public utility divestment, and personal use divestment.

The rapid rise of the fossil fuel divestment movement, the fastest-growing divestment movement in history, is helping to draw attention to the moral urgency of acting on climate change and the broader harms of the fossil fuel industry. A central focus of the movement has been the fact that the fossil fuel industry's collective proven fossil fuel reserves—those already near development—are up to five times greater than can be burned if we are to stand a likely chance of keeping global warming to below 2°C above pre-industrial levels, never mind well below 2°C with the aspiration of hitting 1.5°C as was agreed to under the Paris Agreement (CTI, 2012, 2013; IPCC, 2014; McKibben, 2012). Furthermore, not only do we have too many current reserves, publicly traded fossil fuel companies have also been expending approximately 1 percent of global GDP on developing *new* reserves (CTI,

2013)—tragically, that is about the same amount required to invest in the clean energy economy in order to stay below the 2°C target (cf. IEA, 2014a; Stern, 2007).

If we are to stand a likely chance of keeping warming to 2°C, analysis estimates $28 trillion in lost potential revenue for the fossil fuel industry in the next two decades, creating a potential financial bubble that has come to be referred to as the carbon bubble (Lewis, 2014). Such a loss in revenues would negatively affect not only fossil fuel companies, but potentially those dependent on the fossil fuel industry, such as workers, investors, and fossil fuel-intensive states. If, on the other hand, all the listed fossil fuel reserves are burned, we will drastically overshoot the 2°C limit and instead be on a path to a climate change scenario of up to (if not more than) 4°C by the end of the century, locking us into a future of devastating consequences, including rising sea levels, severe droughts and flooding, food and water shortages, more destructive extreme weather events, widespread species extinction, major increases in human poverty and illness, and large setbacks to economic development (IPCC, 2013; cf. Lynas, 2008; World Bank, 2012).

In the response to the harms that the fossil fuel industry brings, and the contradiction between the fossil fuel industry's collective business model and internationally agreed upon action on climate change, the fossil fuel divestment movement has argued that we have a moral responsibility to divest from fossil fuels. The moral legitimacy of the divestment movement has in turn been challenged by the fossil fuel industry (Ayling, 2017). In this chapter, I aim to defend the moral case for divestment. To help me do so, I begin by turning to an argument for why we are morally obliged to divest from fossil fuels, which was put forward by Katie Ullmann, co-founder of Vanderbilt University's divestment campaign. I aim to point to problems with her argument, not to deny the case for fossil fuel divestment, but rather to strengthen the moral case for divestment, and point to more robust moral arguments in favor of divestment. Below is her argument in full:

1. Things that cause an increase in human suffering and death are bad.
2. Resource scarcity, pollution, and natural disasters cause human suffering and death.
3. The burning of fossil fuels cause resource scarcity, pollution, and natural disasters.
4. The burning of fossil fuels cause human suffering and death.
5. The burning of fossil fuels is bad.
6. If it is in our power to prevent something bad from happening, without thereby sacrificing anything of moral importance, we ought, morally, to do it.

7. We can prevent something bad from happening by divesting from fossil fuels.
8. Divesting from fossil fuels and reinvesting in high-returns clean energy investments does not sacrifice anything of moral importance.
9. In the case of divesting from fossil fuels, we can prevent something bad from happening without sacrificing anything of moral importance.
10. We ought, morally, to divest from fossil fuels (Ullmann, 2013).

While Ullmann's argument might have a certain force for those who have seen climate stability undermined by the often corrupt and deceptive profit-seeking actions of the fossil fuel industry, it does not provide a sound case for divestment, as it is based on several problematic premises. The first problematic premise is premise 3, which holds that the burning of fossil fuels causes resource scarcity, pollution, and natural disasters. While this may be true when looking at the cumulative effects of burning fossil fuels, it is not the case that simply any individual act of burning of fossil fuels in itself causes harm. As such, we need to refine this statement to better understand when the burning of fossil fuels is morally problematic. Relatedly, premises 5, 7, and 8, which claim that fossil fuels are bad without any qualifications, are problematic as they focus only on the detrimental impacts of fossil fuels without recognizing any of the benefits they bring. Doing so leads to the claim that divesting from fossil fuels does not sacrifice anything of moral importance whatsoever, which ignores the fact that fossil fuels provide much of moral significance despite the harm they do also cause. To shore up these problematic premises, Section 8.2 will argue that burning fossil fuels in excess of the Paris Agreement targets is morally problematic because it creates grave, substantial, and unnecessary harm.

Building on the argument in Section 8.2, in Section 8.3 I will argue the moral case for divestment from the fossil fuel industry, based on the moral argument that investing in the fossil fuel industry contributes to grave and unnecessary harm and injustice, and is thus morally wrong. In Section 8.4, I supplement this moral argument not to contribute to harm, with a positive moral argument that agents have positive moral duties to promote climate action. My argument augments and improves upon premise 9 of Ullmann's argument, which states that divesting from fossil fuels can prevent something bad from happening. The difference in my position is that I do not claim that any one divestment will typically by itself stop something bad from happening, but divestment may help prevent harm through promoting broader climate action. Finally, in Section 8.5, I argue for a non-consequentialist argument for divestment, which holds that investing in fossil fuels diminishes the integrity and morally tarnishes those who do so by making them complicit in the injustices of the fossil fuel industry.

8.2. The Unnecessary, Grave, and Substantial Harms of Fossil Fuels

Despite all the harms that fossil fuels do cause, for many, fossil fuels are certainly of some moral importance. Consider, for instance, that coal and other fossil fuels provide jobs, energy, or revenue sources to millions of people across the world. By unlocking vast amounts of energy, fossil fuels have played a key role in the industrial revolution along with the progress (and regress) that came with it. As Cheryl Hall (2013) points out, any trajectory we choose as a society inevitably involves some sacrifice and moral trade-offs, and it is no different with regard to leaving behind fossil fuels for a clean energy future. Thus, Ullman's argument as it stands does not provide sound support for divestment, as it relies on the problematic claim that the fossil fuel industry does not have any moral significance whatsoever. Addressing this point is important, for fossil fuel industry propaganda has caricatured the divestment movement as naïve environmentalists who do not understand the importance of fossil fuels or recognize what it takes to create a prospering society. For example, fossil fuel lobbyist Alex Epstein, in *The Moral Case for Fossil Fuels* (2013), argues that divesting would hinder human development, which is provided in large part by the fossil fuel economy. He concludes that we are morally obligated to use more fossil fuels because of their contribution to prosperity.

While Epstein does a lot to play up the moral importance of fossil fuels, his misguided, misinformed, and arguably intentionally deceitful piece, like that of many fossil fuel apologists, fails to adequately countenance either the harms of fossil fuels or the much better future that can be provided by clean energy and climate action. Thus, while Ullman's argument is problematic for not finding any moral value in fossil fuels, Epstein's argument, like that of many other fossil fuel apologists, is problematic for failing to adequately recognize the immense harms of fossil fuels and the moral value of an alternative clean energy future.

One can find a position between that of Ullman and Epstein, and recognize both that the fossil fuel industry does have some moral importance because of the contributions it makes to current livelihoods and the energy it provides, and that a clean energy future is far preferable, as the limited short-term benefits of staying invested in fossil fuel energy production are significantly outweighed by their costs when compared to the benefits of a clean energy future. To flesh out such a position we can turn to numerous studies which have illustrated that by phasing out fossil fuels in line with the Paris Agreement we can avoid significant amounts of environmental and climate change impacts, while increasing health benefits, economic savings, and job creation. For instance, International Energy Agency (IEA) estimates show that transitioning in line with the Paris Agreement could result in net savings on fuel costs of $71 trillion by 2050. Across Africa and India, IEA estimates suggests that high clean energy penetration in line with the 2°C target could result in significant cost savings and higher energy access in both the near and long term (Calitz et al., 2015; CTI, 2014a; IEA, 2014a; Lenferna, 2016). In order to finance a clean energy future, we could also consider that globally eliminating fossil fuel subsidies could free up $2.9 trillion in government revenue annually (Clements et al., 2013). That is almost twice the estimated annual investment needed in clean energy and energy efficiency by 2035 to meet the Paris Agreement targets of staying within 2°C above pre-industrial levels (IEA, 2014b). If all fossil fuel subsidies were reinvested in a low-carbon future, we might even be able to keep open the rapidly closing window of opportunity to meet the much safer target of staying below 1.5°C (Kuramochi et al., 2016).

Acting in line with the 1.5°C target would bring about major benefits and avoid major harms, many of which fall disproportionately on the global south and global poor (Seager, 2009; Abeysinghe and Huq, 2016). Rogelj et al. (2015) provided economic estimates showing that that even when one excludes consideration of the economic benefits from, for example, avoided climate damages, reduced air pollution and improved energy security, meeting the 1.5°C target would only result in a reduction of a few tenths of a percentage point in global GDP growth per year. When you incorporate those broader benefits, research from the Low Carbon Monitor (2016) shows that action in line with the 1.5°C target would result in major benefits, including creating many more jobs, improved global health, and improved energy access compared both to business-as-usual or the 2°C target. Furthermore, achieving the 1.5°C target rather than the 2°C goal has "enormous repercussions," such as avoiding the virtual disappearance of coral reefs; preventing a 10–15 percent increased risk of crop yield losses for key breadbasket areas in just the coming decades; and averting a 10 percent reduction of the global economy by 2050.

More broadly, the IPCC (2014), the Deep Decarbonization Pathways Project, and other global analyses have shown that "deeply reducing greenhouse gas emissions and achieving socio-economic development are not mutually exclusive. [Rather] robust economic growth and rising prosperity are consistent with the objective of deep decarbonization. They form two sides of the same coin and must be pursued together as part of sustainable development" (Deep Decarbonization Pathways Project, 2014, p. vii). A growing number of studies show that a clean energy future has clear advantages in terms of clean air, water, jobs, energy security, economic growth, and a range of other benefits, such that climate action aligns with our collective interests, regionally, nationally, and internationally (cf. Brown, 2009; IPCC, 2014; Jacobson and Delucchi, 2011).

Additionally, far from fossil fuels being the salve to energy poverty, with prices dropping at a swift rate, clean energy is increasingly much better placed to address energy poverty, as a growing body of evidence demonstrates (Bradshaw, 2017; CTI, 2014b; Kyte, 2015; Lenferna, 2016). Combined with the fact that the harms of climate change and fossil fuels fall disproportionately on low-income communities, indigenous peoples, people of color, women, and the global south, a commitment to prioritizing the poor and to racial, global, and gender justice, favors climate action—points highlighted by the Black Lives Matter movement in their platform's call for fossil fuel divestment, Pope Francis in his encyclical on climate change, and a vast body of academic literature (Abeysinghe and Huq, 2016; Cuomo, 2011; Pope Francis, 2015; Movement for Black Lives, 2016).

It is, however, important not to deny the importance of ensuring that climate action adequately addresses the needs of the poor and marginalized and does not disproportionately impact them. While acting in line with the Paris Agreement can bring about major benefits, it is important not to forget the negative impacts that moving away from fossil fuels might bring, especially to those most dependent on them, such as fossil fuel workers. As such, a just transition which adequately accounts for and addresses the negative impacts of a transition is important, particularly if we want to avoid unnecessary harms in the transition away from fossil fuels. Similarly, we should not forget the dual responsibility that the developed world has to both reduce their emissions and to assist the developing world in transitioning to a resilient low-carbon future (cf. Holz et al., 2017).

Considering the need for a just transition though, what the range of studies highlighted above show is that we can align the global community with the Paris Agreement, while receiving significant benefits through doing so. Recognizing this, we can strengthen Ullman's argument for divestment by replacing premise 6, and connected premises which argue that fossil fuels have no moral value, with arguments that moving away from fossil fuels fulfills our duty not to harm others unnecessarily. Consider, for instance, a broadly supported moral principle that we should avoid doing harm where possible. In its most uncontroversial form, as stated by Elizabeth Cripps, a harm avoidance principle states that: moral agents have a "moral duty to avoid inflicting serious harm... on another human being or human beings... *at least* if she can avoid doing so without suffering comparable harm herself" (Cripps, 2013, p. 11). If we then take it as established that reduction in fossil fuel dependence in line with the Paris Agreement will lead to the avoidance of much harm, then by the *harm avoidance principle* we collectively have the moral responsibility to act in line with the Paris Agreement and in doing so avoid inflicting or contributing to serious harm.

Some might object that harm is an unfortunate but necessary feature of running our societies. For instance, Holly Lawford-Smith argues that private duties to reduce emissions cannot be grounded in an unqualified duty to do no harm as "it is virtually impossible in our current social context—for those in developed countries at least—to do no harm, and we cannot have duties to do what we cannot do" (Lawford-Smith, 2016). She argues that this shifts the general injunction from "do no harm" to "do the least harm." Recognizing Lawford-Smith's point, fortunately, does not undercut the moral case for acting in alignment with the Paris Agreement, because we can refine our moral argument to doing the least harm, or to avoiding unnecessary harm. Drawing on the evidence highlighted earlier, we can argue that acting in line with the fossil fuel industry business model creates unnecessary harms, and that we have a moral responsibility to act in line with at least the Paris Agreement, as doing so can provide us with a more prosperous future without causing significant harm to others. To formalize this point, one can draw on an argument put forward by John Nolt who provides us with a principle, which I will call the *no unnecessary harm principle,* which holds that "one may not contribute significantly and unnecessarily to the bodily harm of others" (Nolt, 2013, p. 140). As Nolt points out, such a principle is supported by most, if not all, widely recognized moral theories, thus giving it significant traction as a firm and widely supported foundation upon which to base the moral imperative to transition away from fossil fuels.

Furthermore, the *no unnecessary harm principle* should be augmented to reflect the fact that it is not just small amounts of harm that would be inflicted if we failed to rapidly transition to a clean energy future, but rather substantial and grave harm. As a report by Cleveland Cutler has succinctly summarized, "the fossil energy system causes pervasive human health, environmental, and social harm across every society" (Cleveland, 2015). It is particularly important to point this out, because some institutions have declined divestment partly on the grounds that the harm caused by fossil fuels is not sufficiently grave (cf. Paxson, 2013). However, if causing disastrous climate change and detrimentally affecting the well-being of billions of people, particularly the poor, marginalized, and vulnerable, is not sufficiently grave, I am not sure what is. Causing the premature deaths of millions through air and water pollution (Landrigan et al., 2017; World Health Organization, 2014), greatly increasing the likelihood of a mass extinction (cf. Nolt, 2011), radically and detrimentally altering human civilization as we know it (cf. Ahmed, 2014; Anderson and Bows, 2011), creating a future defined by catastrophe (cf. Hartzell-Nichols, 2014), contributing to major increases in war and conflict (cf. Burke et al., 2009), and the many other impacts of prolonged reliance on fossil fuels, all seem like very grave harms indeed. Thus, recognizing that continued fossil fuel dependence along the lines of the fossil fuel industry business model creates a grave, unnecessary, and substantial harm provides strong moral reasons not to continue fossil fuel dependence in line with the current business model of the fossil fuel industry, and instead to rapidly transition away from fossil fuels in line with the Paris Agreement, and as close as possible to 1.5°C given the major benefits and avoided harms of doing so.

8.3. Contributing to Harm and Injustice

Section 8.2 provided an argument for why we should collectively transition away from fossil fuels, as acting in line with the fossil fuel industry's business model creates grave, substantial, and unnecessary harm. Building from this argument, we can now turn to arguing the moral case why individuals should divest from fossil fuels. We can flesh out what is wrong with continuing investments in the fossil fuel industry in terms of the role that an agent invested in fossil fuels has in contributing to the harms and injustices of the fossil fuel industry. This can be grounded in an insight put forward by Christian Barry and David Wiens (2016), which argues that seemingly innocent beneficiaries of injustices do wrong to the sufferers of injustice by sustaining and contributing to wrongful harm even if they are not themselves the ones inflicting those harms and are only benefiting from them. According to Barry and Wiens we have relatively stringent moral requirements not to contribute to and sustain wrongful harm, and it seems that such a moral duty applies strongly in this case too. Those that invest in and thereby support the fossil fuel industry are contributing to and sustaining wrongful harm, particularly if they are investing in fossil fuel companies whose actions run contrary to the Paris Climate Agreement targets, and thus entail unnecessary, substantial, and grave harms, and the deep injustices that come with failing to address climate change and transition away from a fossil fuel-intensive economy.

Jeremy Moss has somewhat similarly argued that "by investing in the shares of companies that are 'upstream' producers of fossil fuels, an institution is not directly producing emissions but it is knowingly investing in part of the causal chain that leads to the emissions, which in turn increases the likelihood of harms caused by climate change," by providing capital to fossil fuel companies through share purchases (Moss, 2017, p. 418). He argues that based on negative moral duties to avoid contributing to harm that we therefore have a moral duty to divest.[1] However, Moss's argument says that we should divest

[1] Moss labels his account the complicity account, but I think that this is a misuse of the term complicity, as his argument is more about contributing to harm, rather than mere complicity,

from fossil fuels without specifying what constitutes an unacceptable or unjust level of fossil fuel use or harm. Such a broad unspecified claim leaves Moss's view open to Lawford-Smith's previously mentioned objection about the unavoidability of societies causing some harm. Similarly, it leaves Moss's view susceptible to the straw man argument that divestment activists want us to stop using fossil fuels overnight, which is not what most divestment activists are advocating for—rather they are advocating for a rapid and just transition away from fossil fuels.

We can strengthen Moss's view by supplementing it with the account I provided in Section 8.2, about how transitioning away from fossil fuels in line with the Paris Agreement can avoid grave, substantial, and unnecessary harm, and as such we should divest from companies whose actions are out of line with the Paris Agreement. Amending Moss's view accordingly, means it can also be used to support the recent call by the United Nations for investors to align their investments with the Paris Agreement, and also provides a possible moral foundation to underpin the recently proposed "Principles to Guide Investment Toward a Stable Climate," which calls for companies to develop a net-zero emissions business plan in line with keeping warming to either well below 2°C or 1.5°C (Millar et al., 2018).

The case for divesting from the fossil fuel industry is also about more than just divesting from companies whose business models happen to be out of line with the Paris Agreement. The fossil fuel industry is actively lobbying and spreading misinformation to prevent climate action and ensure that their deeply harmful business model is enacted. As extensive research has revealed, fossil fuel companies and utilities are often actively undermining public understanding on climate change and spending billions of dollars pushing back against policies which would allow us to move beyond fossil fuels (Conway and Oreskes, 2010; McKinnon, 2016). The coordinated campaign of misinformation and lobbying against progress on climate change may be one of the most egregious corporate crimes ever committed as it stalls urgently needed action on climate change and clean energy, thus undermining climate stability and endangering the lives of billions. While, especially those of us in the developed world, all contribute to the harms of climate change through our emissions, what is particularly egregious about the fossil fuel industry's contribution to climate injustice and thus particularly problematic about investing in them, is that the fossil fuel industry is not just meeting supply and demand, as Shell's CEO and many others claim, rather they are actively promoting inaction (cf. Leggett, 2016, p. 27). Thus, those investing in the

which seems weaker, as you can be complicit in a crime without actively contributing to it. As such, I refer to his account as the negative harm and/or contributory argument and discuss complicity in my later non-consequentialist argument.

fossil fuel industry would be supporting and implicitly endorsing efforts to actively undermine climate progress.

Following the precedent of tobacco divestment, we could thus divest on the principle of not contributing financially to companies who spread harmful misinformation and corrupt effective public policymaking. For higher education institutions such as my own, the University of Washington, this would be a particularly pertinent principle given that its primary mission is the "preservation, enhancement and dissemination of knowledge." As such investments in and support of fossil fuel industry misinformation would typically run against its primary mission and that of many higher education institutions. Additionally, institutions with their own particular moral commitments could appeal to those values too. For instance, we could also add another specific principle regarding human rights violations. This is relevant not only because the burning of fossil fuels and climate change leads to significant human rights violations (cf. Caney, 2010), but also because of the fossil fuel industry's long record of more direct indigenous and human rights violations, such as Shell's horrific history in the Niger Delta or Exxon's human rights abuses in Indonesia (Bond, 2011). Additionally, a religious or environmental institution could incorporate their own commitment to act as stewards of creation and the natural world, as the World Council of Churches did when it decided to divest (Vaughan, 2014). Combining these points shows that we have a moral responsibility to divest, particularly from companies that are out of line with the Paris Agreement, for otherwise we are contributing to grave, substantial, and unnecessary harm, as well as potentially to misinformation and corruption, human rights violations, and the undermining of ecological well-being.

8.4. Promoting Climate Justice Through Divestment

Apart from negative moral duties not to contribute to unnecessary harm and injustice, we can also justify the moral case for fossil fuel divestment based on the positive moral responsibilities we have to help promote climate action. These duties are referred to as positive duties to help prevent harm, as they are distinguished from negative moral duties, which are moral duties not to cause harm, of the sort highlighted in Section 8.3.[2] In the case of divestment, both positive and negative moral duties are arguably at play, as those invested in

[2] In the philosophical literature, duties to stop causing or contributing to harm are taken to typically be more stringent moral duties than positive moral duties to help prevent harm (Pogge, 2005).

fossil fuels are both contributing to harm through their investments, and through divesting they can also fulfill broader positive moral responsibilities to help promote collective action—Elizabeth Cripps (2013) refers to the latter as "promotional duties," which are moral duties to promote the sort of collective action needed to tackle climate change.

As Marion Hourdequin (2011) argues, overcoming collective action problems requires the actions of many individuals acting in concert, both to reduce their own relative impacts and contributions to the problem, and to help promote broader collective action. Promotional duties are particularly important in the face of a complex structural injustice such as climate change, which requires strong moral leadership to counteract the structures and entrenched interests upholding climate inaction. By exposing and stigmatizing the problematic role of the fossil fuel industry and highlighting the urgency of the carbon bubble, divestment can play (and arguably is already playing) an important role in helping promote broader collective action on climate change. Elsewhere, I have more fully evaluated the empirical evidence supporting the efficacy of divestment (Lenferna, 2018). As that paper highlights, there is a growing body of literature which demonstrates divestment's efficacy in shifting moral, political, cultural, and financial norms, and in shifting capital out of the fossil fuel industry and into clean energy (Ansar et al., 2013; Grady-Benson and Sarathy, 2016; Rowe et al., 2016; Yona and Lenferna, 2016).

As critics are quick to point out, any one divestment by itself will be unlikely to prevent harm occurring. To a certain extent, the critics are right, but they miss the point. Like many actions aimed at addressing climate change, a single action, like a single divestment, is unlikely to cause or prevent harm by itself, although some bank divestments may be able to prevent harm directly if by refusing to finance a particular fossil fuel project they prevent the project from going ahead. More broadly though, like most collective action problems, the more appropriate question is not whether a single divestment commitment alone can prevent harm, but rather whether investing or divesting can collectively contribute to or prevent harm from occurring. To borrow the words of Derek Parfit it is a "mistake in moral mathematics" to assume that because an act has an imperceptible effect or makes only a tiny contribution to a cumulative harm, that it therefore cannot be wrong (Parfit, 1984, p. 77). This is because, as Chris Cuomo points out, "if one knows her actions are part of a set of collective actions that together result in great harm, she must evaluate the rightness or wrongness of her contributions in light of the knowledge that others are also engaging in the activity, and together they create a cumulative effect" (Cuomo, 2011). Furthermore, an additional point which pushes back against the idea that divestment is insignificant is to recognize that while relatively unknown and non-wealthy individuals taking divestment actions will have smaller impacts, typically the divestment movement has focused on larger, more influential institutions whose influence is much more pronounced.

Some have argued, however, that we should not divest if it puts considerable financial burdens on institutions and/or prevents other ways of acting on climate change and may interfere with other promotional duties. Of course, we need to be careful about divestment drawing too much attention from other actions, especially those that might be more effective. Indeed, we must be mindful, as many divestment activists are, of the fact that divestment by itself is far from sufficient to meet the Paris Agreement, and that we will need much broader action. However, divestment may not necessarily weaken other climate actions, and may rather empower them (Rowe et al., 2016). Furthermore, there is a growing body of evidence that one can divest without reduced financial performance, and may actually perform better financially and protect itself from losses associated with an industry potentially in decline (Trinks et al., 2018).[3] As such, divestment may fulfill moral responsibilities both to promote action and stop contributing to unjust harm, while also providing stronger financial returns and enabling broader action.

8.5. A Non-Consequentialist Moral Case for Divestment

The moral case for divestment in the previous arguments relies at least partly on the ability for divestment to have a broader impact on reducing fossil fuel use and preventing harm. Thus, it provides somewhat of a consequentialist account of the moral case for divestment. What makes it consequentialist is that it focuses on the ability for divestment to make a difference in the world by contributing to or reducing harm—the consequences are what is morally important. For those who remain skeptical of divestment's impact, or perhaps of consequentialist moral theory in general, let us now bracket out the possible impacts that divestment might have. If we do so, can we still make a moral case for divestment?

One non-consequentialist argument in favor of divestment relies on the virtue of integrity. As Marion Hourdequin points out, "the kind of unity that integrity recommends requires that an individual work to harmonise her commitments at various levels and achieve a life in which her commitments

[3] If one does not grant that fossil fuel divestment actually increases returns, another way of responding to concerns that divesting from fossil fuel may place undue burdens on some, is to acknowledge that duties to act on climate change are relative to our capabilities and positions in society. To support this, one can turn to Robin Eckersley (2014) and Eric Godoy (2017) who both argue, drawing on the work of Iris Marion Young (2011), that the extent of one's responsibility in the face of structural injustices like climate change is relative to one's power, privilege, interest, and collective ability—to that list I would also add that our responsibilities are also relative to the extent to which one has contributed to and benefited from the problem.

are embodied not only in a single sphere, but in the various spheres she inhabits" (Hourdequin, 2010, p. 448). Such unity is simply not available if an agent or institution claims to be committed to climate justice while investing in climate failure. On this line of reasoning, the many institutions that have committed to the goals of the Paris Agreement and yet invest contrary to them, are guilty of a lack of integrity.

This lack of integrity also contributes to a moral tarnishing of the non-divested agent. As Elizabeth Cripps argues, "those individuals who can, but choose not to, distance themselves from groups that are contributing to harm are metaphysically guilty, and also morally responsible for, in the sense of being morally tainted by, the harms caused by their fellow group members or institutions" (2013, p. 177). By investing their financial resources into fossil fuel companies, institutions are complicit and supportive of the industry's injustices and harm, and are morally tainted as a result. Even if an individual's actions do not increase the likelihood that the harm will occur, which arguably they do, nonetheless they would be complicit and endorse the potential grave, substantial, and unnecessary harms of the fossil fuel industry through their investment, and this tarnishes the moral character of the agent.

As Henry Shue points out (in a comment on this chapter), moral tarnishing is both non-consequentialist, as it has to do with concerns about the sort of character an agent has, but it is also simultaneously consequentialist, as by endorsing such activities agents contribute to the broader perception of them as acceptable. By divesting, on the other hand, one distances oneself from the fossil fuel industry, and increases the growing stigma attached to the grave, substantial, and unnecessary harms the industry is collectively planning on causing. Stigmatization of the fossil fuel industry has been a powerful element of the fossil fuel divestment movement, as stigmatization detracts from the legitimacy of the fossil fuel industry, and legitimacy is crucial to their continued acceptance and support in society, such that undermining legitimacy paves the way for needed action on climate change (Ayling, 2017; Rowe et al., 2016).

8.6. Conclusion

In conclusion, there is a moral case for fossil fuel divestment grounded in both consequentialist and non-consequentialist moral reasoning. Given this, if one wants to deny the moral case for divestment one would have to provide an alternative course of action which could compellingly discharge that moral responsibility or provide an objection which undermines the multiple grounds I have provided in support of a moral case for divestment. The consequentialist moral case is grounded both in promotional duties to act on climate change, and the responsibility to not contribute to the potential unnecessary, grave, and substantial harms associated with the fossil fuel industry's continued business model. The non-consequentialist moral case asks institutions to act with integrity and avoid the moral tarnish that comes from investing in, and thus being supportive of and complicit in, the injustices and grave harms entailed in the fossil fuel industry business model. By divesting, institutions and individuals can play an important leadership role by acting on the deep moral urgency of the climate crisis, moving their investments out of a substantial, grave, and unnecessarily harmful fossil fueled future, and investing instead in a prosperous, low-carbon future, which prevents the worst ravages of climate change and fossil fuel dependency, at least while it is still possible to do so.

Acknowledgments

Many thanks are due to the divestment movement who have collectively been a rich resource of wisdom for this chapter. For more direct input on this chapter and related research, I am greatly thankful to Stephen Gardiner, Henry Shue, Lauren Hartzell-Nichols, Brett Fleishman, Ann Cudd, Bruce Herbert, Augustin Fragnière, Alec Connon, Morgan Sinclaire, John Nolt, Leehi Yona, Brian Henning, members of the University of Washington Philosophy Department, Divest UW, 350 Seattle, and audiences at the 2016 Climate Justice Conference at Cornell University, the 2014 Chicago Central APA, Gonzaga University, Western Washington University, the University of Washington, and many more.

References

Abeysinghe, A., and S. Huq. 2016. "Climate Justice for LDCs through Global Decisions." In *Climate Justice in a Non-Ideal World*, edited by D. Roser and C. Heyward, 189–207. Oxford: Oxford University Press.

Ahmed, N. 2014. "Nasa-Funded Study: Industrial Civilisation Headed for 'Irreversible Collapse'?" *The Guardian*. <http://www.theguardian.com/environment/earth-insight/2014/mar/14/nasa-civilisation-irreversible-collapse-study-scientists>.

Anderson, K., and A. Bows. 2011. "Beyond 'Dangerous' Climate Change: Emission Scenarios for a New World." *Philosophical Transactions of the Royal Society A* 369: 20–44.

Ansar, A., B. Caldecott, and J. Tilbury. 2013. "Stranded Assets and the Fossil Fuel Divestment Campaign: What Does Divestment Mean for the Valuation of Fossil Fuel Assets?" <http://www.bsg.ox.ac.uk/stranded-assets-and-fossil-fuel-divestment-campaign-what-does-divestment-mean-valuation-fossil-fuel>.

Ayling, J. 2017. "A Contest for Legitimacy: The Divestment Movement and the Fossil Fuel Industry." *Law and Policy* (accepted article). <http://onlinelibrary.wiley.com/doi/10.1111/lapo.12087/abstract>.

Barry, C., and Wiens, D. 2016. "Benefiting from Wrongdoing and Sustaining Wrongful Harm." *Journal of Moral Philosophy* 13 (5): 530–52.

Bond, P. 2011. *Politics of Climate Justice: Paralysis Above, Movement Below*. Scottsville: University of KwaZulu-Natal Press.

Bradshaw, S. 2017. "More Coal Equals More Poverty: Transforming Our World Through Renewable Energy." <https://www.oxfam.org.au/wp-content/uploads/2017/05/More-Coal-Equals-More-Poverty.pdf>.

Brown, L. R. 2009. *Plan B 4.0*. New York: Earth Policy Institute. <http://www.earth-policy.org/images/uploads/book_files/pb4book.pdf>.

Burke, M. B., E. Miguel, S. Satyanath, J. A. Dykema, and D. B. Lobell. 2009. "Warming Increases the Risk of Civil War in Africa." *Proceedings of the National Academy of Sciences of the United States of America* 106 (49): 20670–4.

Calitz, J., C. Mushwana, T. and Bischof-Niemz. 2015. "Financial Benefits of Renewables in South Africa in 2015: Actual Diesel- and Coal-fuel Savings and Avoided 'Unserved Energy' from the First Operational 1.8 GW of Wind and PV Projects in a Constrained South African Power System." <http://www.csir.co.za/media_releases/docs/Financial benefits of Wind and PV 2015.pdf>.

Caney, S. 2010. "Cosmopolitan Justice, Responsibility and Global Climate Change." In *Climate Ethics: Essential Readings*, edited by S. M. Gardiner, S. Caney, H. Shue, and D. Jamieson, 122–45. New York: Oxford University Press.

Clements, M., D. Coady, and M. Fabrizio. 2013. "Energy Subsidy Reform: Lessons and Implications." International Monetary Fund (iii): 68. <https://doi.org/http://dx.doi.org/10.5089/9781475558111.071>.

Cleveland, C. J. 2015. "Why Divest? The Substantial Harm of Fossil Fuels." *Energy in Context*, March 2. <http://energyincontext.com/2015/03/why-divest-the-substantial-harm-of-fossil-fuels/>.

Conway, E. M., and N. Oreskes. 2010. *Merchants of Doubt*. New York: Bloomsbury Press.

Cripps, E. 2013. *Climate Change and the Moral Agent: Individual Duties in an Interdependent World*. Oxford: Oxford University Press.

CTI. 2012. "Unburnable Carbon—Are the World's Financial Markets Carrying a Carbon Bubble?" <http://www.carbontracker.org/wp-content/uploads/downloads/2012/08/Unburnable-Carbon-Full1.pdf>.

CTI. 2013. "Unburnable Carbon 2013: Wasted Capital and Stranded Assets." <http://www.carbontracker.org/wastedcapital>.

CTI. 2014a. "Energy Access: A Guide to Why Coal is Not the Way Out of Poverty." <http://www.carbontracker.org/wp-content/uploads/2014/11/Coal-Energy-Access-111014-final.pdf>.

CTI. 2014b. "Energy Access: Why Coal is Not the Way Out of Energy Poverty." <http://www.carbontracker.org/wp-content/uploads/2014/11/Coal-Energy-Access-111014-final.pdf>.

Cuomo, C. J. 2011. "Climate Change, Vulnerability and Responsibility." *Hypatia* 26 (4): 690–711.

Deep Decarbonization Pathways Project. 2014. "Pathways to Deep Decarbonization." <http://unsdsn.org/wp-content/uploads/2014/09/DDPP_Digit.pdf>.

Eckersley, R. 2014. "Responsibility for Structural Injustices: The Case of Climate Change." University of Queensland Research Seminar. <http://www.polsis.uq.edu.au/responsibility-structural-injustices-climate>.

Epstein, A. 2013. "The Moral Case for Fossil Fuels." <http://industrialprogress.com/moralcase/>.

Godoy, E. S. 2017. "Sharing Responsibility to Divest from Fossil Fuels." *Environmental Values* 26 (6): 693–710.

Grady-Benson, J., and B. Sarathy. 2016. "Fossil Fuel Divestment in US Higher Education: Student-Led Organising for Climate Justice." *Local Environment* 21 (6): 661–81.

Hall, C. 2013. "What Will it Mean to be Green? Envisioning Positive Possibilities Without Dismissing Loss." *Ethics, Policy & Environment* 16 (2): 125–41.

Hartzell-Nichols, L. 2014. "Adaptation as Precaution." *Environmental Values* 23: 149–64.

Holz, C., S. Kartha, and T. Athanasiou. 2017. "Fairly Sharing 1.5: National Fair Shares of a 1.5°C-compliant Global Mitigation Effort." *International Environmental Agreements: Politics, Law and Economics*. <https://doi.org/10.1007/s10784–017–9371-z>.

Hourdequin, M. 2010. "Climate, Collective Action and Individual Ethical Obligations." *Environmental Values* 19 (4): 443–64.

Hourdequin, M. 2011. "Climate Change and Individual Responsibility: A Reply to Johnson." *Environmental Values* 20 (2): 157–62.

IEA. 2014a. *Energy Technology Perspectives 2014: Harnessing Electricity's Potential*. Paris: IEA. <http://www.iea.org/etp/>.

IEA. 2014b. *World Energy Outlook 2014*. Paris: IEA. <http://www.iea.org/publications/freepublications/publication/WEO_2014_ES_English_WEB.pdf>.

IPCC. 2013. "AR5." <http://www.ipcc.ch/report/ar5/wg1/#.UlWi71Csim4>.

IPCC. 2014. "Summary for Policymakers". In *Climate Change 2014: Mitigation of Climate Change. Contribution of Working Group III to the Fifth Assessment Report of the Intergovernmental Panel on Climate Change*. <http://mitigation2014.org/>.

Jacobson, M. Z., and M. A. Delucchi. 2011. "Providing All Global Energy with Wind, Water, and Solar Power, Part I: Technologies, Energy Resources, Quantities and Areas of Infrastructure, and Materials." *Energy Policy* 39 (3): 1154–69.

Kuramochi, T. et al. 2016. "10 Steps: The Ten Most Important Short-Term Steps to Limit Warming to 1.5°C." <http://www.ecofys.com/files/files/climate-action-tracker-2016–10-steps-for-1-point-5-goal.pdf>.

Kyte, R. 2015. "World Bank: Clean Energy is the Solution to Poverty, Not Coal." *The Guardian*, August 10. <https://www.theguardian.com/sustainable-business/2015/aug/07/world-bank-clean-energy-is-the-solution-to-poverty-not-coal>.

Landrigan, P. J. et al. 2017. "The Lancet Commission on Pollution and Health." *The Lancet* 6736 (17).

Lawford-Smith, H. 2016. "Climate Matters Pro Tanto, Does it Matter All-Things-Considered?" *Midwest Studies in Philosophy* 40 (1): 129–42.

Leggett, J. 2016. "The Winning of the Carbon War: Power and Politics on the Front Lines of Climate and Clean Energy." Colophon. <http://www.jeremyleggett.net/download-page/>.

Lenferna, G. A. 2016. "How Africa Could Leapfrog Fossil Fuels to Clean Energy Alternatives." *The Conversation*, March 1. <https://theconversation.com/how-africa-could-leapfrog-fossil-fuels-to-clean-energy-alternatives-55044>.

Lenferna, G. A. 2018. "Divestment as Climate Justice: Weighing the Power of the Fossil Fuel Divestment Movement." In *Climate Justice and the Economy: Social Mobilization, Knowledge and the Political*, edited by S. G. Jacobsen. Routledge Earthscan. <https://www.academia.edu/22236651/Divestment_as_Climate_Justice_Weighing_the_Power_of_the_Fossil_Fuel_Divestment_Movement>.

Lewis, M. C. 2014. "Stranded Assets, Fossilised Revenues." <http://www.keplercheuvreux.com/pdf/research/EG_EG_253208.pdf>.

Low Carbon Monitor. 2016. "Pursuing the 1.5°C Limit: Benefits and Opportunities." <http://www.thecvf.org/wp-content/uploads/low-carbon-monitor-lowres.pdf>.

Lynas, M. 2008. *Six Degrees: Our Future on a Hotter Planet.* Washington, DC: National Geographic.

McKibben, B. 2012. "Global Warming's Terrifying New Math." *Rolling Stone.* <http://www.rollingstone.com/politics/news/global-warmings-terrifying-new-math-20120719>.

McKinnon, C. 2016. "Should We Tolerate Climate Change Denial ?" *Midwest Studies in Philosophy* 40: 205–16.

Millar, R. J., C. Hepburn, J. Beddington, and M. R. Allen. 2018. "Principles to Guide Investment Towards a Stable Climate." *Nature Climate Change* 8 (January): 1–3.

Moss, J. 2017. "The Morality of Divestment." *Law and Policy* (accepted article): 1–39.

Movement for Black Lives. 2016. "Invest-Divest." <https://policy.m4bl.org/invest-divest/>.

Nolt, J. 2011. "Nonanthropocentric Climate Ethics." *WIRES Climate Change* 2: 701–11.

Nolt, J. 2013. "The Individual's Obligation to Relinquish Unnecessary Greenhouse-Gas-Emitting Devices." *Philosophy and Public Issues* 3 (1): 139–65.

Parfit, D. 1984. *Reasons and Persons*. Oxford: Clarendon Press.

Paxson, C. H. 2013. "Coal Divestment Update." <http://brown.edu/about/administration/president/2013-10-27-coal-divestment-update>.

Pogge, T. 2005. "Severe Poverty as a Violation of Negative Duties." *Ethics & International Affairs* 19 (1): 55–83.

Pope Francis. 2015. *Laudato Si: On Care for Our Common Home.* Vatican: The Word Among Us Press.

Rogelj, J., G. Luderer, R. C. Pietzcker, E. Kriegler, M. Schaeffer, V. Krey, and K. Riahi. 2015. "Energy System Transformations for Limiting End-of-Century Warming to Below 1.5°C." *Nature Climate Change* 5: 519–27. <http://www.nature.com/nclimate/journal/v5/n6/full/nclimate2572.html>.

Rowe, J., J. Dempsey, and P. Gibbs. 2016. "The Power of Fossil Fuel Divestment (and its Secret)." In *A World to Win: Contemporary Social Movements and Counter-Hegemony*, edited by W. K. Carroll and K. Sarker. Winnipeg: ARP Books.

Seager, J. 2009. "Death by Degrees: Taking a Feminist Hard Look at the 2 Degree Climate Policy." *Women, Gender & Research* 18 (3–4): 11–21.

Stern, N. 2007. *The Economics of Climate Change: The Stern Review*. Cambridge: Cambridge University Press.

Trinks, A., B. Scholtens, M. Mulder, and L. Dam. 2018. "Fossil Fuel Divestment and Portfolio Performance." *Ecological Economics* 146 (July 2017): 740–8.

Ullmann, K. 2013. "Simply, We Ought to Divest." <http://www.insidevandy.com/opinion/article_21f3fd92-a3af-11e2-be92–0019bb30f31a.html>.

Vaughan, A. 2014. "World Council of Churches Rules Out Fossil Fuel Investments." *The Guardian*, July 11. <http://www.theguardian.com/environment/2014/jul/11/world-council-of-churches-pulls-fossil-fuel-investments>.

World Bank. 2012. *Turn Down the Heat: Why a 4 Degree Celsius World Must Be Avoided.* Washington, DC: World Bank. <http://climatechange.worldbank.org/sites/default/files/Turn_Down_the_heat_Why_a_4_degree_centrigrade_warmer_world_must_be_avoided.pdf>.

World Health Organization. 2014. "7 Million Premature Deaths Annually Linked to Air Pollution." <http://www.who.int/mediacentre/news/releases/2014/air-pollution/en/>.

Yona, L., and A. Lenferna. 2016. "The Fossil Fuel Divestment Movement Within Universities." In *Environment, Climate Change and International Relations*, edited by G. Sosa-Nunez and E. Atkins. E-International Relations, <http://www.e-ir.info/2016/05/15/the-fossil-fuel-divestment-movement-within-universities/>.

Young, I. M. 2011. *Responsibility for Justice*. New York: Oxford University Press.

9

Justice and Posterity

Simon Caney

Humanity faces a number of serious challenges. Notwithstanding increases in the standard of living there is still extensive global poverty and inequality. In addition to this, however, economic growth conducted along the lines of business-as-usual threatens to cause dangerous climate change, as well as to cross other "planetary boundaries" (Rockström et al., 2009; Steffen et al., 2015). Given this, we need to consider how to balance the demands of the present generation, in particular the most vulnerable and disadvantaged on this planet, with those of future generations. To do this we need a theory of justice that includes both current and future people in its scope.

This chapter provides a sympathetic exploration of one conception of intergenerational justice.[1] Stated very roughly, it holds that the members of one generation should act in such a way that they leave future people with a standard of living that is at least equal to their own. Different versions of this principle have been advanced by political philosophers, economists, and international lawyers. This underlying idea is an intuitively appealing one and worth exploring in more detail. At the same time, the principle is quite elusive,

[1] I will argue later that it is a mistake to think of our responsibilities to future people in terms of duties to future "generations" (Section 9.2). For this reason it is somewhat misleading to refer, as I do here and elsewhere in this chapter, to "intergenerational justice." (I also think it is unhelpful for two other reasons. First, as I have argued elsewhere (Caney 2018a), the term "generation" is often defined in several quite distinct, and often inconsistent, ways, and so the term "intergenerational justice" can be confusing. In addition to this, the term "intergenerational justice" is potentially inaccurate in another sense for I (like others) am using it to refer to duties of justice owed to future people, but one might reasonably think that intergenerational justice includes both duties of justice owed by one generation to past generations, as well as those owed to future generations.)

A more accurate term than "intergenerational justice" would be "justice to future people" (Section 9.4). However, this is also more cumbersome and so I will sometimes use the term "intergenerational justice" as a shorthand.

and we can fruitfully distinguish between different ways of interpreting the core idea. In addition to this, it is not always made clear why we should adopt this principle and what follows from it. The aims of this chapter are therefore threefold:

1. to distinguish between different versions of this principle and address several unclarities (Sections 9.2–9.5);
2. to explore what might be said in its defense (Section 9.6); and
3. to explore what follows from this principle, and, in particular, what it implies for addressing two great evils which afflict our age—global poverty and climate change (Sections 9.7–9.8).

9.1. The Core Idea

Let us start then with the core idea. Many have suggested that current generations should leave future generations with a standard of living that is no worse than the one that they themselves enjoy. For example, at the end of his seminal essay "Justice Between Generations," Brian Barry writes that "those alive at any time are custodians rather than owners of the planet, and ought to pass it on in *at least no worse shape than they found it in*" (Barry, 1991b [1977], p. 258, italics added). Justice, in his view, precludes leaving future generations worse off: we must leave a world that is at least as good as ours. But, he thinks, we may leave them better off. Elsewhere he endorses the view that there is a distinction "between making successors better off, which is a nice thing to do but not required by justice, and not making them worse off, which *is* required by justice" (Barry, 1991a [1979], p. 241; see also pp. 240–1).[2] James Woodward (who cites Barry as a source) has a nice formulation of the central idea. He writes that "each generation ought to leave for succeeding generations a total range of resources and opportunities which are at least equal to its own range of resources and opportunities" (Woodward, 1986, p. 819).

The same core idea is also expressed by Tim Mulgan in his book *Future Persons* (2006). Mulgan assents to what he terms "[t]he No Worse-off Principle," which he defines as follows: "[n]o generation should leave later generations worse-off than itself, if it can avoid this" (Mulgan, 2006, p. 197). He expands on this, writing that "it is clearly wrong for one generation gratuitously to leave later generations worse off by unnecessarily depleting resources" (Mulgan, 2006, p. 198).

[2] See also Barry (1978, 243–4; 1991c [1983], especially pp. 259ff.).

Edith Brown Weiss affirms a recognizably similar principle of intergenerational justice. In her seminal text on intergenerational justice and international law she writes that "each generation is entitled to inherit a planet and cultural resource base at least as good as that of previous generations" (Brown Weiss, 1989, p. 25).

If we turn now to the discipline of economics, we can find another statement of the same broad idea in the work of the Nobel laureate in economics, Robert Solow. In his *An Almost Practical Step Toward Sustainability*, Solow writes that "[t]he duty imposed by sustainability is to bequeath to posterity not any particular thing—with the sort of rare exception I have mentioned—but rather to endow them with whatever it takes to achieve a standard of living at least as good as our own and to look after their next generation similarly" (Solow 1992, 15).[3]

Now these formulations differ in a number of respects. In what follows I take the core idea (or at least the version that I am most sympathetic to) to have the following structure:

> *The Principle of "Justice to Future People" (JFP)*: X (roughly, those alive at one time) should leave others, Y (those born in the future), *no worse off than themselves.*

This core principle is, clearly, susceptible to a variety of different interpretations and we need to adjudicate between competing conceptions of the underlying concept.[4] In the next four sections I discuss four questions that need to be addressed before we have a clear understanding of the central principle at stake, and the motivating idea. In answering these questions I explore the implications of the answers reached for climate change policy.

9.2. What Are the Moral Units of Justice?

The first question can be introduced as follows. As already noted, the Principle states that some, X, should leave others, Y, no worse off than themselves, but to what kinds of entities do X and Y refer?

Many write as if the entities in question are "generations." Mulgan's formulation, for example, states that "[n]o *generation* should leave later *generations* worse-off than itself, if it can avoid this" (Mulgan, 2006, p. 197, emphasis

[3] The "rare exception[s]" that Solow has in mind are "certain unique and irreplaceable assets [that] should be preserved for their own sake" (Solow, 1992, p. 14). A closely related principle is defended by Llavador et al. in their *Sustainability for a Warming Planet* (2015), as well as by Roemer in single-authored papers (see Roemer, 2011, esp. pp. 377–9; and 2013, pp. 145–7, esp. p. 146).

[4] For the distinction between a concept and a conception see Rawls (1999, p. 5).

added). Interpreted literally this treats "generations" as the fundamental duty-bearers and rights-bearers. Similarly, Woodward writes that "each *generation* ought to leave for succeeding *generations* a total range of resources and opportunities which are at least equal to its own range of resources and opportunities" (Woodward, 1986, p. 819, emphasis added). This also treats each "generation" as the relevant duty-bearer and asserts that the duties are owed to "*succeeding* generations" (my emphasis). Brown Weiss (1989, p. 25) too treats generations as the relevant rights-holders.[5]

In line with my earlier work on global justice and intergenerational justice, I believe that we should *not* treat "generations" as the fundamental rights- or duty-bearers (Caney, 2005a, 2009). Rather, I submit, we should adopt a disaggregated approach.[6] On this approach, the relevant fundamental moral units (such as individuals) are often subsets within any given generation and may also include members of more than one generation.

§1. Consider first the duty-bearer. It seems to me unhelpful to treat "generations" as the fundamental duty-bearers. Suppose we consider the question: who is responsible for ensuring a fair response to the prospect of dangerous climate change—one which both effectively prevents dangerous climatic changes and which imposes the burdens fairly? We might give two different kinds of answer. Many (myself included) will appeal to what have been termed "first-order duties." This refers to duties on the part of individuals and others to engage in (or abstain from) some acts—in this case, it is a duty to mitigate climate change and not to emit more than one's fair share of greenhouse gases. Second, and in addition to that, there are also "second-order duties" to act in ways that incentivize agents to comply with their first-order responsibilities and discourage noncompliance.[7]

Now in both cases, the most plausible principles entail that the responsibilities would not be held by a generation *en masse*, but rather by a group of agents which both are a subset of any generation and which also cross generational boundaries. In the case of first-order responsibilities, duties to mitigate will plausibly vary in line with factors such as an agent's level of wealth or vulnerability and their access to sources of energy other than fossil fuels (Caney, 2005b, 2006, 2010, 2018b; Shue, 2014). Given this, the extent to

[5] I should add that I am not claiming that these thinkers necessarily think that "generations" are indeed the fundamental moral units. It is much more likely that at least some of them (in particular Mulgan and Woodward) are using "generations" simply as a shorthand, and that their concern is with individual people.

[6] For a distinct, but illuminating, discussion of the importance of disaggregating (current and future) generations in discussions of intertemporal justice, see Thomas Schelling's discussion of disaggregation in his analysis of the social discount rate: Schelling (1995, pp. 398–400; 2000, pp. 835–6).

[7] For the distinction between first- and second-order responsibilities see O'Neill (2005, pp. 428, 433–6). For my use of it in the context of climate change, see Caney (2014a, pp. 134–6).

which an agent has a duty to mitigate will vary considerably among the members of any generation. It would, for example, be implausible to treat struggling Indian peasants in the same way as an affluent citizen of the US. A similar conclusion holds for second-order responsibilities. As I have argued elsewhere, the most plausible principle holds that those with the greatest political power bear the greatest responsibility (Caney, 2014a, pp. 141ff.). Again then this calls for both (i) differentiating within any given generation and also (ii) ascribing responsibilities to members from different generations.

§2. Consider now the rightful recipients. Again, it is misleading to adopt an aggregative approach and to treat "generations" as the fundamental rights-holders. For this would entail that the duty-bearers at t1 should act so that a later generation or generations has a certain standard of living, and it would be indifferent to the distribution of benefits and burdens within any future generation. After all, on this "aggregative" view, justice is said to be owed "generations."

Such a view is, however, implausible. In the first place, a "generation," however, defined, is a somewhat arbitrary and artificial unit that does not have any intrinsic value. Ontologically it is also odd to divide the future of humanity into separate and discrete chunks—a generation 1, generation 2, and so on—rather than see a flow of future people.

Second, to conceive of people's duties as being owed to the collective unit that is a generation (or a number of generations) comes with a moral cost for it obscures from view the distribution of wealth and income *within* any future generation. To illustrate the point, let us suppose that the designated duty-bearers in one or more generations can affect *both* the standard of living enjoyed by the members of a later generation, G_t, and the distribution of wealth and income within G_t. Suppose then that we consider what obligations these duty-bearers will have to the members of G_t. Imagine that they are then presented with the following two policy options:

Policy 1: this policy

(a) generates considerable benefits for future people in generation G_t (so advances their absolute standard of living); and

(b) creates and entrenches great inequality in G_t.

Policy 2: this policy

(a) generates benefits for future people in G_t but it leaves them slightly less well off than under policy 1; and

(b) creates an equal society in G_t.

Faced with such a choice, the best choice might be to opt for *Policy 2* (depending on the absolute standard of living under *Policy 2*, how great the differences in the standard of living are between *Policies 1* and *2*, and the magnitude of the inequalities in *Policy 1*). Why? The answer is that we care not simply about the absolute standard of living but also the distribution. Given this, duty-bearers have reason to be concerned with distributions within future generations.

To put the point another way: we might say, following Rawls, that there is a natural duty to support just institutions (Rawls, 1999, pp. 99ff.), and that by selecting *Policy 1* and bequeathing a deeply unequal society to future people in G_t, the duty-bearers in the preceding generations will have violated that.[8] Those who are committed to an egalitarian ideal of justice cannot ignore this.

Now one response might be that the duty to bring about a just distribution of income and wealth within G_t falls to the members of G_t alone. But this is unrealistic for it fails to recognize the role and importance of path-dependence and historically entrenched inequalities. The policies adopted by one generation can leave a distribution of capital skewed in favor of a minority, and they can bequeath a political system that is unresponsive to the needs of the poor and vulnerable. These can be very difficult—perhaps impossible—to undo. To use Charles Tilly's term (1998), one generation can create "durable inequalities" and it is naïve to think that future citizens can readily undo the inequitable economic and political system that they inherit.

In short, it is a mistake to think that those duty-bound to uphold principles of intertemporal justice owe duties to "future generations." The phrasing is at best misleading, suggesting that there are duties to discrete groupings of people—a generation 1, generation 2, etc.—and it wrongly hides from view the responsibilities that those at one point in time may have with respect to the distribution of wealth and income *within* any future generations. Rather than say that duties are owed to "future generations," it may be better to think of them as being owed to "future people."[9]

[8] Rawls himself says that his principle of intergenerational justice ("the savings principle") is "an interpretation . . . of the . . . natural duty to uphold and to further just institutions" (1999, p. 257). Rawls often describes the requirement of intergenerational justice in rather vague terms, referring to furthering "just institutions" (1999, p. 257). However, in one paragraph he does seem to suggest that it requires people at one point in time to further the realization of the Difference Principle in future societies (1999, p. 258). This would naturally follow from this starting point.

[9] Should we drop all reference to "generations" altogether and treat all future people as an undifferentiated whole? No, not necessarily. The fact that two people are contemporaries can have some derivative moral significance. Suppose that A has more wealth than B. If they were contemporaries this might lead to B feeling stigmatized and might undermine B's sense of self-respect (O'Neill, 2008, pp. 121–3; Scanlon, 2018, ch. 3). However, if A and B are not contemporaries then neither of these are likely to occur (the present do not feel stigmatized by the anticipated wealth of subsequent generations). Inequalities of wealth and income between contemporaries can thus have a significance that inequalities of wealth and income between non-contemporaries lacks. Indeed, my earlier argument drew attention to the significance of such inequalities in wealth and income among one generation. So generational membership may have this derivative significance. It is, though, wholly derivative. Our focus should still be on the condition of individuals (whatever generation), and *economic* inequalities in one generation have significance only because and to the extent that they causally affect people's standard of living.

9.3. What is the Intertemporal Scope of Justice?

Consider now a second question. Let us suppose that (as I have suggested in Section 9.2) when it comes to specifying the entities to whom duties of intergenerational justice are owed the right answer is "individuals." The next question that arises is "to which individuals do they owe duties?" Are their responsibilities to the members of the next generation? Or, to those in the next two or three generations? Or, to all who will ever live in the future? Some write as if the duties are owed to the members of the next generation. Others employ vague phrases like "succeeding generations" (Woodward, 1986, p. 819) and "later generations" (Mulgan, 2006, p. 197). The question here then is: what is the temporal scope of an agent's responsibilities?

My proposal here is that we adopt what I shall term the *Causal Impact Principle*. This holds that the duties of those alive at one time extend as far into the future as the effects of their actions (and inactions). To illustrate, consider the use of nuclear energy. The waste created is expected to have an impact for 10,000 years (Archer, 2009, p. 11). Or consider the emission of greenhouse gases. Archer reports that if we burn a ton of coal, then 25 percent of the CO_2 emitted will be exerting an impact on the climate system in 1,000 years and 10 percent will be doing so in 100,000 years (Archer, 2009, p. 1). In the age of the Anthropocene the scope of justice can then extend very far into the future.

Two points should be noted about this proposal. The first concerns how we frame the duty bearing in mind that the duty-bearers may have a flawed or even completely mistaken (but entirely reasonable) understanding of how far into the future they will leave a mark. Suppose that *A*'s actions have impacts for 300 years, but that he could not have reasonably been expected to know that. It seems implausible to hold him to be accountable to all whose standard of living is affected by his actions. Given the possibility of ignorance, it is best to frame agents' responsibilities in terms of what they can reasonably be expected to know given the evidence that they have at their disposal.[10]

[10] See Derek Parfit's discussion of "evidence-relative" notions of wrongness (and the contrast with "fact-relative" and "belief-relative" notions) (Parfit, 2011, pp. 150–1). The statement in the text is not quite correct as it stands for it should be supplemented with an additional point, namely that we should not take the existing level of evidence as a given. Duty-bearers might have insufficient evidence because they have been negligent and have failed to undertake the appropriate amount of research. (Suppose that someone implements a radical new technology without doing any research on the possible side effects. Given their failure to do research they can say (correctly) that there is no evidence that it would be harmful. But this is clearly not good enough. So even if their actions are in line with the evidence they have, that does not mean that their action is justified.) They thus also have a duty to undertake research so that, other things being equal, they have the best understanding that they could have of their impacts into the future.

Second, it is important to distinguish between, on the one hand, who is duty-bound to take a decision such as mitigating climate change, and, on the other hand, who should bear the burden of that decision. The Causal Impact Principle entails that some of those alive now are under a *duty* to all future people for the next 1,000 years to effect the transition to a zero carbon economy. However, to make this claim about the scope of the duty-bearers' responsibilities is *not* to say that they must necessarily bear the entirety of the burden involved in doing so. Just because *A* owes a duty to *B* to do *y* does not entail that *A* should pay for the entirety of the cost of doing *y*. If, for example, they can pass some of the cost of doing so on to affluent people who live in the future (Broome, 2012, ch. 3, esp. p. 44; Rendall, 2011, esp. pp. 892–3) then that might be permissible, depending on what duties we have to future people.[11]

9.4. What is the Currency of Justice?

We turn now to a third question. Any statement to the effect that those alive at one period of time should leave future people no worse off than them needs to be able to provide an answer to the question "no worse off *in terms of what*?" That is, what are the criteria by which we define some people as being "worse off" than others (or "better off" than them or "equally well off" to them)? To use the philosophical terminology, our third question is: what is the "currency" of (intertemporal) justice (Cohen, 1989)?

The answer we give will have considerable implications for future people. Suppose, for example, that we think that justice should be concerned with fair shares to *natural resources*, then this considerably restricts what individuals are entitled to do during their lifetime. It would, for example, disallow the members of one generation destroying a natural resource but leaving a substitute in its place. As Barry (1991c [1983], esp. pp. 259–60) and Solow (1992, pp. 14–15) note, this is unduly restrictive. Why should we privilege *natural* resources if a human-created substitute can provide exactly the benefits that a natural resource can without any additional harmful effects?

In light of this we might adopt a broader account of the currency of justice and define it simply as *resources* (where this is understood to include not only natural resources but also physical capital, knowledge, and expertise) (Barry, 1991c [1983], pp. 260ff.; Woodward, 1986, p. 819).

However, this too faces a problem. For as Amartya Sen argued long ago, focusing on resources is a kind of "fetishism" (Sen, 1980, p. 216). Sen's point is that resources are merely a means to an end and do not have value in and of

[11] For discussion see Caney (2014a, 2014b).

themselves. We value resources because of what they enable us to do and achieve. We care, the argument goes, about what Sen (2009, part III) and Martha Nussbaum (2006) have termed "capabilities." In light of this, I adopt the "capabilities" approach that has been pioneered and developed with considerable sophistication by Nussbaum and Sen. I draw, in particular, on Nussbaum's instructive specification of capabilities. The capabilities, on this approach, include "life," "bodily health," "bodily integrity," "senses, imagination, and thought," "emotions," "practical reason," "affiliation," "other species," "play," and "control over one's environment" (Nussbaum, 2006, pp. 76–8). Persons' responsibilities to the future thus require leaving future people a fair share of these capabilities.[12]

From this point of view, it is a serious mistake to evaluate the impacts of climate change (or any other kind of phenomenon, such as ocean acidification or biodiversity loss) solely for their impact on GDP or resources. Rather, temperature increases, severe weather events, and sea-level rises should be judged according to their impact on a broader range of human (and non-human)[13] interests, including their attachments to land and existing communities.[14]

9.5. What is the Baseline?

We now know what criteria to employ to judge whether one agent is worse off or not than another. However, the statement that those alive now should not leave future people worse off than *themselves* is still somewhat unclear. Whose standard of living (understood as a set of capabilities) should we use as a baseline by which to determine the entitlements of future people?

§1. *Non-moralized baselines*. One suggestion might be to select the standard of living enjoyed by some people today (call it level S) and then hold that those alive now should leave future people no worse off than this level S. This, though, runs into some obvious problems. The first is: whose standard of living would one select—that of a contemporary US citizen? Or a Portuguese citizen? Or Italian?—and on what basis would one choose theirs and not that of others? Furthermore, if one chose one level (say that of the average contemporary US citizen) it might simply be unsustainable to maintain that into

[12] Of course, a great deal more could be said about other potential accounts of the currency of justice (such as 'utility') but space precludes discussing the other accounts here.

[13] I take it that duties of justice are owed not simply to human beings but also non-human animals, but shall not attempt to defend that here.

[14] For a discussion of the impacts on climate change on cultural communities see Heyward (forthcoming).

the future. Alternatively, if one chose another level (say, that of a contemporary Nigerian citizen) then, while this could be sustained into the future, it is vulnerable to the objection that it is too low and too undemanding a standard. Justice to contemporaries and to future people, one might very reasonably argue, requires more than that.

More generally, it is hard to see why we should adopt any non-moralized state of affairs (such as the way the world is now) as the baseline, and tie the claims of future people to this level. It assumes—without argument—that the status quo has some kind of moral legitimacy.

§2. *Moralized baselines*. Given this, a much more promising line would be to adopt an explicitly normative criterion. With this in mind, my proposal is this: we should begin with our understanding of what justice between contemporaries requires. On my view, a just world is one in which all have equal capabilities to function (Caney, 2005a). Bearing this in mind we should then seek to identify *what would be the highest level at which the system of equal capabilities could be enjoyed by those alive at one time such that it would also leave future people with an at least equally good set of capabilities*. I submit that this—what one might term the Maximal Equal Standard of Living criterion—is how we should think of the baseline.

A similar (but distinct) view is defended by Llavador et al. in their important book *Sustainability for a Warming Planet* (2015, pp. 1–5 and 34–8). They too emphasize maximizing the equal rights of the present subject to leaving future people equally well off. At this point, however, their view and mine diverge. Mine *allows* people to leave future people better off than they are, but does not require it. They, however, require the state to leave future people better off. They build in what they term "growth sustainability"—that is, a commitment to promote utility by a certain percentage each year (Llavador et al., 2015, pp. 4ff. and 35ff.; Roemer, 2013, p. 146).

9.6. Normative Considerations

The preceding four sections have sought to clarify the underlying principle of justice, and, in doing so, to present the most appealing version. This, however, is a natural point at which to ask: why should we use the mooted distributive principle at all? Why are we required to leave future people no worse off?

I think the best way to motivate support for it is to compare with some of the competing principles.

§1. *Sufficiency*. Some, for instance, adopt a sufficientarian approach, and hold that everyone should be above a decent threshold standard of living (Page, 2006, ch. 4, esp. pp. 90–5). This is surely a necessary feature of our responsibilities to the future. However, it would be implausible to claim that it is sufficient.[15] The claim that *all* that is owed to future people is that they be above a certain threshold would, for example, permit a state of affairs in which those alive at one point in time hand themselves luxurious standards of living but leave future people with no more than the minimum. Such a policy would penalize some for no reason other than that they are born in the future rather than now, and this (their date of birth) is a dubious reason for allocating them poorer opportunities. Future people could object: "there is no morally relevant difference between you and us, so how could it be acceptable to deny us the same opportunities that you extend to yourselves?" The only difference is their temporal location, but this lacks any intrinsic importance and is not a good reason for granting some worse opportunities than others (Caney, 2009, pp. 168–9; 2014b, pp. 323–4).

§2. *Equality*. In light of this, one might think that we should just adopt a strictly egalitarian approach. This, however, is vulnerable to the leveling down objection (Page, 2006, pp. 80–2; Parfit, 1997, pp. 210–1 and 218–20). For example, could we really insist on equality if we could leave future people better off at no extra cost to those alive now? Suppose, for example, we can design a clean technology that will benefit current generations but will leave future people even better off. Must we—in the name of intergenerational equality—stem the flow of benefits? That seems very unappealing.

This brings us neatly to the view I am defending. For like the egalitarian ideal it does not discriminate against future people. In addition to this, it is compatible with leaving future people better off and so does not stand condemned by the leveling down objection.

Clearly, more can, and should, be said in defense of this principle. Space precludes discussing it further here. I hope though to have motivated some support for the principle under consideration.

9.7. A Restatement

At this point it may be useful to provide a summary of where we have got to so far. I have made four proposals about how best to construe the Principle of Justice for Future People.

First, I have proposed that we should not conceive of the duties affirmed by it as being owed by "generations" to "generations." Rather they are owed by some (duty-bearers) who are likely to be both a subset of one generation and to

[15] For an excellent discussion of this see Casal (2007).

include members of more than one generation. Furthermore, the duty is owed to "future people"—not "future generations"—and should be concerned with distributions within generations as well as across them.

Second, we face the question: which future people? How far into the future does the duty extend? I have proposed that those alive at one time can be held accountable for the impacts of their actions as far into the future as they can reasonably be expected to foresee.

My third proposal was that the duty to leave people at least equally well off should employ a capabilities account to define what is meant by "equally well off."

Finally, to the question of "What baseline should we employ?" I have suggested that we employ what I termed the Maximal Equal Standard of Living. Duty-bearers should, then, seek to create a world in which all enjoy an equal standard of living (defined in terms of the enjoyment of capabilities) and that they should do so at the highest level possible consistent with leaving future people with an at least equally good standard of living (again defined in terms of the enjoyment of the capabilities).

Drawing on the above and my work elsewhere (Caney, 2005a), we can summarize the view as follows. The guiding principle is:

> Principle I. There is a duty on the part of the powerful and affluent to ensure that current people enjoy the maximal equal standard of living that is consistent with leaving future people with a standard of living that is at least as good.

This requires at least three kinds of duties to future people. First:

> Principle II. There is a duty to promote the absolute standard of living of future people as far into the future as one can reasonably be expected to foresee so that they enjoy a standard of living that is at least as good (the Absolute Component).

In addition to this, my earlier argument in Section 9.2 (calling for the disaggregation the rights holder) leads to a second duty to future people:

> Principle III. There is a duty to promote the realization of equality *within* future generations (the Distribution Sensitive Component).

Principle III is important and is routinely neglected in discussions of intergenerational equity.

In the interests of providing a comprehensive account, we should also acknowledge a further duty to future people, namely

> Principle IV. There is a duty to act in such a way that it leaves those who live in the future well placed to discharge their duties to those who come after them.

We wrong people if we make it unreasonably onerous for them to discharge their duties, and we also act irresponsibly with respect to the beneficiaries of those duties if we needlessly make it less likely that those duty-bound to them will comply with their responsibilities.

9.8. From Theory to Practice

The next question to ask is: what follows from this account? What implications does it have for issues like development and for climate change?

I will answer this by way of setting out six points.

§1. The first point is that the Maximal Equal Standard of Living would clearly call for the eradication of global poverty. It is committed to raising the starving, malnourished, the ill, and the impoverished up to a standard of living such that we can guarantee it for future as well as current people.

§2. Second, realizing Principle II requires an ambitious mitigation program. But what should the specific target be? For example, should our aim be to keep the increase in global mean temperatures to no more than 1.5°C over pre-industrial times? Or is a 2°C limit sufficient? Any target will of course rest on empirical analyses of the impacts of climate change on people's lives, but it also rests on normative assumptions about what is owed to the potential victims of climate change and what is owed to those who will bear the burden of any climate policy. My principle provides an ethical compass that can inform (though not determine) what the appropriate goal ought to be and how any burdens should be shared between people over time. A just climate treaty must be such that it optimizes the equal standard of living of current people while also leaving future people at least as well off.

§3. Third, it is worth noting that my Principle III gives us an additional—less familiar—reason for mitigating climate change. This principle, recall, maintains that people have a *pro tanto* duty not to act in ways that create, worsen, or entrench intragenerational inequalities in the future. Now, as the IPCC has often reported, climate change is expected to exacerbate future inequalities (Olsson et al., 2014). Furthermore, more recently, research has found that climatic effects will also emerge more quickly in the low-lying areas that are disproportionately inhabited by the world's least advantaged (Harrington et al., 2016)—thereby aggravating existing inequalities. Thus, Principle III, like Principle II, also calls for an ambitious mitigation program.

§4. My account yields a fourth point, namely that given the commitment to global equality it is vital that any burden involved in mitigating climate change and thus honoring duties to future people does not come at the cost of inequality now. My point about disaggregation (Section II) is thus highly relevant here. Statements to the effect that the "current" generation must bear burdens for the benefit of future generations are misleading here, for the egalitarian component of my first principle would call for the least advantaged not to be burdened and for the burdens to be borne by the most affluent (Caney, 2010). They would thus endorse proposals to target the top billion emitters (Chakravarty et al., 2009).[16]

§5. Furthermore, to realize our responsibilities to the future without compromising the rights of the poor to develop will also require clean technology transfer. In fact, what it calls for is a more egalitarian system of the ownership of clean technology, and a shift from it being monopolized by a minority to a world in which access is shared. Here it is instructive to note the finding by Hauke Ward et al. that transferring existing technology would lower annual CO_2 emissions by more than 10 gigatons (Ward et al., 2017). Taking a step toward greater equality, and the common ownership of humanity's technological capacities, also helps to realize responsibilities to the future and to create a more just and sustainable future.

§6. Some may fear that my account of responsibilities to the future imposes unreasonable burdens. I will therefore close with three observations. First, such concerns overlook the extent to which honoring duties to future people (such as mitigating climate change), in practice, can be done without cost and with co-benefits, such as enhanced energy security and improved air quality (Edenhofer et al., 2014, p. 62).

Second, they also overlook the fact that doing justice to contemporaries can contribute to doing justice to future people. Several reasons why this might be the case have been noted already. An additional reason for thinking this arises from the role that population change plays in putting pressures on the climate system and from the effects of promoting global justice on demographic change. To explain: population growth is now recognized to be a driver of climate change (Edenhofer et al., 2014, p. 47). We thus have reason to seek to dampen world population growth. It is therefore highly relevant that many policies that are required by duties of justice to the world's most disadvantaged further this goal. For example, John Bongaarts reports that providing access to affordable contraception reduces the fertility rate in developing countries, drawing on evidence from Bangladesh, Iran, and Rwanda where extending real access to contraception has led to dramatic reductions in the number of children per woman (Bongaarts, 2016). Sarah Bradley and her co-authors find that fully satisfying the unmet demand for contraception in developing countries would reduce the Total Fertility Rate by 20 percent (Bradley et al., 2012, 57).[17] Wolfgang Lutz and Samir KC place a special

[16] For further discussion of points §4–§6 see Caney (2018c).

[17] As Abel and his co-authors note, this figure reflects *existing demand* for contraception, and the increased empowerment of women can be expected to lead to a reduction in demand and thus a further reduction in the Total Fertility Rate (Abel et al., 2016, p. 14296).

emphasis on education, arguing that it plays a crucial role in depressing the fertility rate (Lutz and KC, 2011). More generally, Abel et al. find that meeting the Sustainable Development Goals "would result in the world population peaking around 2060 and reaching 8.2–8.7 billion by 2100" (Abel et al., 2016, p. 14294).

A third point also bears noting. To realize responsibilities to future people we need to have a clear understanding of the determinants of a well-functioning society, one equipped to realize the maximal equal standard of living in the future and over time. Now so far I have emphasized the *ecological* determinants. It is important to recognize, however, that the other central determinants include well-functioning political institutions. The extensive literature on economic development emphasizes the centrality of inclusive institutions and the long-run impacts of those institutions (Acemoglu et al., 2001), and it chronicles how the unjust imposition of extractive institutions can lead to a massive "reversal of fortune" and immiseration for centuries to come (Acemoglu et al., 2002). In short, unjust institutions wreak havoc for many years, centuries even. Related to this, political scientists have shown the pivotal importance of institutions characterized by impartiality and a lack of corruption (Rothstein, 2011). Justice to future people involves bequeathing these. Doing so, however, does not require major additional economic burdens or large transfers of wealth across time. It would thus be a mistake to think of the responsibilities here requiring considerable economic sacrifice.

9.9. Conclusion

Equality is a powerful ideal. Many, however, have taken it to apply only to contemporaries. My suggestion in this chapter is that we should move beyond such a restrictive account of the scope of justice, and should instead include in its remit not just those alive now but those who will be born. Our goal, I have argued, should be to create a world in which those alive now enjoy the highest equal standard of living that is consistent with leaving future people at least as well off. Anything less short-changes them and denies their equal moral status. Realizing this will require tackling global inequalities and the environmental degradation that disfigures our planet, and it will require reforming political institutions so that they are less presentist and so that they serve the interests of humanity.

Acknowledgment

I am very grateful to Henry Shue for his comments and suggestions.

References

Abel, Guy J., Bilal Barakat, Samir KC, and Wolfgang Lutz. 2016. "Meeting the Sustainable Development Goals Leads to Lower World Population Growth." *Proceedings of the National Academy of Sciences* 113 (50): 14294–9.

Acemoglu, Daron, Simon Johnson, and James A. Robinson. 2001. "The Colonial Origins of Comparative Development: An Empirical Investigation." *American Economic Review* 91 (5): 1369–401.

Acemoglu, Daron, Simon Johnson, and James A. Robinson. 2002. "Reversal of Fortune: Geography and Institutions in the Making of the Modern World Income Distribution." *Quarterly Journal of Economics* 117 (4): 1231–94.

Archer, David. 2009. *The Long Thaw: How Humans are Changing the Next 100,000 Years of Earth's Climate*. Princeton, NJ and Oxford: Princeton University Press.

Barry, Brian. 1978. "Circumstances of Justice and Future Generations." In *Obligations to Future Generations*, edited by R. I. Sikora and Brian Barry, 204–48. Philadelphia: Temple University Press.

Barry, Brian. 1991a [1979]. "Justice as Reciprocity." In *Liberty and Justice: Essays in Political Theory Volume 2*, 211–41. Oxford: Clarendon Press.

Barry, Brian. 1991b [1977]. "Justice Between Generations." In *Liberty and Justice: Essays in Political Theory Volume 2*, 242–58. Oxford: Clarendon Press.

Barry, Brian. 1991c [1983]. "The Ethics of Resource Depletion." In *Liberty and Justice: Essays in Political Theory Volume 2*, 259–73. Oxford: Clarendon Press.

Bongaarts, John. 2016. "Development: Slow Down Population Growth." *Nature* 530 (7591): 409–12.

Bradley, Sarah E. K., Trevor N. Croft, Joy D. Fishel, and Charles F. Westoff. 2012. *Revising Unmet Need for Family Planning*. DHS Analytical Studies No. 25. Calverton, MD: ICF International.

Broome, John. 2012. *Climate Matters: Ethics in a Warming World*. New York and London: W. W. Norton.

Brown Weiss, Edith. 1989. *In Fairness to Future Generations: International Law, Common Patrimony, and Intergenerational Equity*. New York: Transnational Publishers.

Caney, Simon. 2005a. *Justice Beyond Borders: A Global Political Theory*. Oxford: Oxford University Press.

Caney, Simon. 2005b. "Cosmopolitan Justice, Responsibility, and Global Climate Change." *Leiden Journal of International Law* 18 (4): 747–75.

Caney, Simon. 2006. "Environmental Degradation, Reparations, and the Moral Significance of History." *Journal of Social Philosophy* 37 (3): 464–82.

Caney, Simon. 2009. "Climate Change and the Future: Discounting for Time, Wealth and Risk." *Journal of Social Philosophy* 40 (2): 163–86.

Caney, Simon. 2010. "Climate Change and the Duties of the Advantaged." *Critical Review of International Social and Political Philosophy* 13 (1): 203–28.

Caney, Simon. 2014a. "Two Kinds of Climate Justice: Avoiding Harm and Sharing Burdens." *Journal of Political Philosophy* 22 (2): 125–49.

Caney, Simon. 2014b. "Climate Change, Intergenerational Equity, and the Social Discount Rate." *Politics, Philosophy & Economics* 13 (4): 320–42.

Caney, Simon. 2018a. "Justice and Future Generations." *Annual Review of Political Science* 21: 475–93.

Caney, Simon. 2018b. "Climate Change and Distributive Justice." In *The Oxford Handbook of Distributive Justice*, edited by Serena Olsaretti, 664–88. Oxford: Oxford University Press.

Caney, Simon. 2018c. "Human Rights, Population and Climate Change." In *Human Rights and 21st Century Challenges: Poverty, Conflict, and the Environment*, edited by Dapo Akande, Jaakko Kuosmanen, Helen McDermott, and Dominic Roser. Oxford: Oxford University Press.

Casal, Paula. 2007. "Why Sufficiency is Not Enough." *Ethics* 117 (2): 296–326.

Chakravarty, Shoibal, Ananth Chikkatur, Heleen de Coninck, Stephen Pacala, Robert Socolow, and Massimo Tavoni. 2009. "Sharing Global CO_2 Emission Reductions Among One Billion High Emitters." *Proceedings of the National Academy of Sciences* 106 (29): 11884–8.

Cohen, G. A. 1989. "On the Currency of Egalitarian Justice." *Ethics* 99 (4): 906–44.

Edenhofer, Ottmar, Ramón Pichs-Madruga, Youba Sokona, Susanne Kadner, Jan C. Minx, and Steffen Brunner. 2014. "Technical Summary." In *Climate Change 2014: Mitigation of Climate Change. Working Group III Contribution to the Fifth Assessment Report of the Intergovernmental Panel on Climate Change*, edited by Ottmar Edenhofer et al., 33–107. Cambridge: Cambridge University Press.

Harrington, Luke, David J. Frame, Erich M. Fischer, Ed Hawkins, Manoj Joshi, and Chris D. Jones. 2016. "Poorest Countries Experience Earlier Anthropogenic Emergence of Daily Temperature Extremes." *Environmental Research Letters*, 11 (5): 055007.

Heyward, Clare. Forthcoming. *The Cultural Dimension of Climate Justice*. Cheltenham: Edward Elgar.

Llavador, Humberto, John E. Roemer, and Joaquim Silvestre. 2015. *Sustainability for a Warming Planet*. Cambridge, MA: Harvard University Press.

Lutz, Wolfgang, and Samir KC. 2011. "Global Human Capital: Integrating Education and Population." *Science* 333 (6042): 587–92.

Mulgan, Tim. 2006. *Future Persons: A Moderate Consequentialist Account of our Obligations to Future Generations*. Oxford: Clarendon Press.

Nussbaum, Martha C. 2006. *Frontiers of Justice: Disability, Nationality, Species Membership*. Cambridge, MA: Harvard University Press.

Olsson, Lennart, Maggie Opondo, and Petra Tschakert. 2014. "Livelihoods and Poverty." In *Climate Change 2014: Impacts, Adaptation, and Vulnerability. Part A: Global and Sectoral Aspects. Contribution of Working Group II to the Fifth Assessment Report of the Intergovernmental Panel on Climate Change*, edited by Christopher B. Field et al., 793–832. Cambridge: Cambridge University Press.

O'Neill, Martin. 2008. "What Should Egalitarians Believe?" *Philosophy and Public Affairs* 36 (2): 119–56.

O'Neill, Onora. 2005. "The Dark Side of Human Rights." *International Affairs*, 81 (2): 427–39.

Page, Edward A. 2006. *Climate Change, Justice and Future Generations*. Cheltenham: Edward Elgar.

Parfit, Derek. 1997. "Equality and Priority." *Ratio* 10 (3): 202–21.

Parfit, Derek. 2011. *On What Matters: Volume One*, edited and introduced by Samuel Scheffler. New York: Oxford University Press.

Rawls, John. 1999. *A Theory of Justice*, revised edition. Oxford: Oxford University Press.

Rendall, Matthew. 2011. "Climate Change and the Threat of Disaster: The Moral Case for Taking Out Insurance at Our Grandchildren's Expense." *Political Studies* 59 (4): 884–99.

Rockström, Johan, Will Steffen, Kevin Noone, Åsa Persson, F. Stuart III Chapin, Eric Lambin, Timothy M. Lenton, Marten Scheffer, Carl Folke, Hans Joachim Schellnhuber, Björn Nykvist, Cynthia A. de Wit, Terry Hughes, Sander van der Leeuw, Henning Rodhe, Sverker Sörlin, Peter K. Snyder, Robert Costanza, Uno Svedin, Malin Falkenmark, Louise Karlberg, Robert W. Corell, Victoria J. Fabry, James Hansen, Brian Walker, Diana Liverman, Katherine Richardson, Paul Crutzen, and Jonathan Foley (2009) "Planetary Boundaries: Exploring the Safe Operating Space for Humanity." *Ecology and Society* 14 (2) (article 32). <http://www.ecologyandsociety.org/vol14/iss2/art32/>.

Roemer, John E. 2011. "The Ethics of Intertemporal Distribution in a Warming Planet." *Environmental and Resource Economics* 48 (3): 363–90.

Roemer, John E. 2013. "Once More on Intergenerational Discounting in Climate-Change Analysis: Reply to Partha Dasgupta." *Environmental and Resource Economics* 56 (1): 141–8.

Rothstein, Bo. 2011. *Quality of Government: Corruption, Social Trust, and Inequality in International Perspective*. Chicago. University of Chicago Press.

Scanlon, T. M. 2018. *Why Does Inequality Matter?* New York: Oxford University Press.

Schelling, Thomas C. 1995. "Intergenerational Discounting." *Energy Policy* 23 (4/5): 395–401.

Schelling, Thomas C. 2000. "Intergenerational and International Discounting." *Risk Analysis* 20 (6): 833–7.

Sen, Amartya. 1980. "Equality of What?" The Tanner Lecture on Human Values. <https://tannerlectures.utah.edu/_documents/a-to-z/s/sen80.pdf>.

Sen, Amartya. 2009. *The Idea of Justice*. London: Allen Lane.

Shue, Henry. 2014. *Climate Justice: Vulnerability and Protection*. Oxford: Oxford University Press.

Solow, Robert. 1992. *An Almost Practical Step Toward Sustainability*. Washington, DC: Resources for the Future.

Steffen, Will, et al. 2015. "Planetary Boundaries: Guiding Human Development on a Changing Planet." *Science* 347 (6223): 1259855.

Tilly, Charles. 1998. *Durable Inequality*. Berkeley and Los Angeles: University of California Press.

Ward, Hauke, Alexander Radebach, Ingmar Vierhaus, Armin Fügenschuh, and Jan Christoph Steckel. 2017. "Reducing Global CO_2 Emissions with the Technologies We Have." *Resource and Energy Economics* 49: 201–17.

Woodward, James. 1986. "The Non-Identity Problem." *Ethics* 96 (4): 804–31.

10

Discounting and the Paradox of the Indefinitely Postponed Splurge

Matthew Rendall

It is sometimes claimed that unless we discounted future benefits, no one would ever get to enjoy them. Tjalling Koopmans spoke of the "paradox of the indefinitely postponed splurge," in which, if cost–benefit analysis gave too much weight to future benefits, their cumulative utility would make it perennially more attractive to reinvest resources than consume them.[1] "[I]f low discount rates are imposed on every generation," warns Robert Mendelsohn, "then every generation is made worse off because they are all burdened by the generations to come."[2] Jan Narveson speaks of "the 'jam tomorrow' paradox: the benefit is always in the future, and each actual generation is miserable."[3]

While the paradox continues to be cited to justify a positive rate of time preference,[4] it has received remarkably little analysis.[5] It should not be confused with the claim that a rate of zero time preference would call for unreasonable sacrifices from the present generation. Many have argued that if we gave the future the same weight as the present, utility maximization would become extremely demanding, since the benefits that investment could bring to vast numbers of our descendants would outweigh nearly any cost to us. Here, in contrast, the objection is that a zero rate of time preference would not even maximize utility. While the paradox of the indefinitely postponed splurge (hereafter PIPS) is less well known than the "argument from excessive sacrifice,"[6] it is in a sense a more compelling objection. Hardened utilitarians might insist that we ought to save at whatever rate maximizes utility, and that

[1] Koopmans, 1967, p. 8. [2] Mendelsohn, 2008, p. 52. [3] Narveson, 1978, p. 59.
[4] Heath, 2017, p. 449.
[5] The longest treatment of which I am aware (Tarsney, 2017) devotes about two pages to the argument, making criticisms that I will extend here.
[6] Parfit, 1987, p. 484.

if that seems unreasonable, our intuitions are likely to be biased, since it is we who would bear the burden. But if nobody would ever enjoy the benefits, then there would not be even a utilitarian case for biting this bullet. Undiscounted utilitarianism (hereafter UU) would be a *self-defeating theory*—a theory that defeats its own aim—here, the maximization of utility.[7] That really would be a *reductio ad absurdum*.

The problem is not limited to utilitarianism. PIPS seems to pose a challenge to any aggregative consequentialist theory that does not ignore the further future. Many egalitarians and all prioritarians, for example, assign greater weight to costs and benefits that affect the worse off.[8] If our descendants are expected to be better off, an egalitarian or prioritarian cost–benefit analysis will give future costs and benefits less weight. Nevertheless, provided that there are *enough* future people who stand to benefit, and their welfare counts for *something*, even these theories, or so it may seem, will judge it preferable for each generation to save all its disposable income than to consume. It may seem that the only way to block this crazy conclusion is to discount future costs and benefits away.

On closer inspection, the argument falls apart. I will assume utilitarianism, but the core argument applies equally, *mutatis mutandis*, to any other consequentialist theory. There is no way a refusal to discount could ensnare us in a self-defeating savings loop. The real problem is indeterminacy: without being able to foresee when humans and other sentient life will go extinct, we have no way to determine what savings rate would maximize utility across generations. That being the case, we need a utilitarian decision-making procedure that stands a decent chance of approximating the right answer. Discounting future costs and benefits could be a legitimate part of this procedure, but only when constrained with a rule—such as Hans Jonas's imperative of responsibility—that gives priority to avoiding catastrophes.

10.1. Is UU Directly Collectively Self-Defeating?

Moral theories, as Derek Parfit explains, can be directly or indirectly self-defeating. A *directly self-defeating theory* is one which would defeat its own purpose if successfully implemented. An *indirectly self-defeating theory* defeats its own purpose if agents *try* to implement it, because they will fail. Theories can also be individually or collectively self-defeating. A *directly individually self-defeating theory* defeats its aim if the agent implements it; an *indirectly individually self-defeating theory*, if she tries to do so. *Directly collectively*

[7] Koopmans, 1967, p. 9. [8] Parfit, 1995/2002.

self-defeating theories defeat their aims if we all implement them, as do *indirectly collectively self-defeating theories* if we all make the attempt.[9] Since PIPS's claim is that we would become trapped in perpetual saving if each generation followed UU, we focus here on collectively self-defeating theories.

Arguments invoking PIPS usually claim that UU is *directly collectively self-defeating*: it *would* lead to perpetual saving if every generation successfully followed the theory. Yet it might seem obvious that no problem could result: if UU's goal is to maximize utility, and everyone successfully maximizes utility, then how could it be self-defeating? It is not quite that simple. We can imagine hypothetical cases in which agents can always increase the payoff by deferring action, as in a case that resembles PIPS:

> *Trust.* You are the trustee of a charitable trust to save starving people. If the world endures forever in its present state, and the principal grows fast enough, then whenever you wind it up, more people could have been saved by dissolving it at a later date. Yet if you delay forever, no one is saved at all.[10]

Trust presents a puzzle for act utilitarianism. Whenever the agent acts, it seems by waiting another day she could have done better. It will not do to say that the agent should let the trust continue for, say, two more years, and then wind it up. Any adequate moral theory must render judgments that are acceptable when applied not only to a particular case, but also to all analogous cases. Moral theories are universalizable: if they make a judgment about a given case, they make the same judgment about any case where the circumstances are relevantly similar. This is because moral judgments depend—"supervene," to use the technical term—on non-moral facts. Thus, as R. M. Hare observes, to say "Jack did just the same as Jim, in just the same circumstances, and they are just the same sort of people, but Jack did what he ought and Jim did what he ought not" would be as senseless as saying "[t]he two figures are exactly the same shape, but one is triangular and the other not."[11]

In two years, the trustee will face just the same set of morally relevant considerations as she does now. If it is right for her to delay two years now, it will also be right for her to delay another two years then, and so on, ad infinitum, with the result that the money will never be spent. But that seems crazy. We can reasonably demand that moral theories should neither be indeterminate, nor defeat their own aims when successfully implemented.[12] Applied to *Trust*, act utilitarianism, or any theory that says it is right to maximize—bring about better consequences rather than worse ones—seems likely to fail this test.

[9] Parfit, 1987, chs. 1–2. [10] Landesman, 1995. [11] Hare, 1961; Hare, 1981, p. 81.
[12] Cf. Parfit, 1987, p. 103.

It might be suggested that the trustee should discount future costs and benefits. Provided her discount rate is high enough, she will dissolve the trust right away and at least do better than by following UU. Unfortunately for the agent, the dilemma in *Trust* can be easily be reformulated so as to take no account of time at all:

Donation. A philanthropist will donate to famine relief whatever sum of money the utilitarian agent names.

No matter what sum she chooses, she could always have picked something higher.[13] The problem arises not because the gains are in the future, but because they are potentially infinite. Infinite values notoriously lead to paradox even when time is not a factor.[14]

While the infinite values in *Trust* and *Donation* pose a problem for utilitarianism, this is not the case for PIPS. At some point the world or the universe will almost surely come to an end, and we will go extinct along with it.[15] Imagine a society of utilitarian archangels, who had perfect knowledge of the future, and always behaved so as to maximize utility.[16] Such a society would have no trouble saving. During the early stages of history they would invest more than they consumed. Gradually, they would reduce investment, and start to dis-save so that the last resources were used up just as life came to an end. There is no reason to think that UU is directly collectively self-defeating.

We need not specify the savings rate that the archangels would select, since *ex hypothesi* they would select whatever was optimific. But it is important to note that they would stop far short of reinvesting all their disposable income. Arguments that invoke PIPS assume that the saving would involve a grim Stakhanovism. It is true that UU would place greater burdens on earlier generations than later ones. But even in the early stages, too narrow a focus on economic development would impair cultural development. Stakhanovism leaves little room for art, music, literature, or any of the rest of the cultural achievements that form much of earlier generations' legacy to the present one.[17] Extreme savings rates can sometimes be maintained for extended periods when populations are convinced or coerced to do so. Stalinism is an example. But this also illustrates the political and cultural impoverishment that can result. Crash industrialization, implemented through widespread terror, not only came at a terrible cost to contemporaries, but also permanently warped Soviet institutions. While it may have saved the USSR from Hitler,[18] nobody could recommend the model as a general strategy for maximizing utility.

[13] Cf. Arntzenius et al., 2004, p. 264. [14] Cowen, 2004, pp. 82–3; Tarsney, 2017, p. 347.
[15] Tarsney, 2017, p. 347. There is some small chance that this assumption is false (Bostrom, 2011), and a correspondingly small chance that PIPS involves the same kind of problem as *Trust*.
[16] Hare, 1981. [17] Revesz, 1999, p. 994. [18] Von Laue, 1981.

The archangels' allocation of investment would differ from Stakhanovism in another way. The marginal utility of material goods declines as societies grow richer, while natural goods such as wilderness increase in value as they become scarcer. Once poverty was overcome, the archangels would channel an increasing share of resources into the preservation of nature.[19] They would do this also for a second reason: as good utilitarians, they would accord great weight to the welfare of wild animals, who are far more numerous than we are. While maximizing utility might require wide-ranging intervention in nature to prevent animal suffering, it seems fair to assume it would also involve strong protection of their habitats.[20] Some natural resources can be hoarded for consumption by future generations, but many such goods—such as wilderness, national parks, and everyday places of natural beauty—cannot. If we preserve these goods for future generations, nothing stops us from making use of some in the meantime, and taking pleasure in the mere knowledge that others exist.

In short, there is no reason to think that UU is directly collectively self-defeating. If every generation successfully acted on the theory, we would start dis-saving in good time and no resources would be wasted. Nor would the lookout even for earlier generations be so grim. The world is already rich enough to eliminate severe poverty if resources were properly distributed. Provided we can avoid disaster, the benefits that we can leave for future generations will increasingly be social, cultural, and natural.

10.2. Is UU Indirectly Collectively Self-Defeating?

We are not archangels. We cannot reliably foresee the end of the world and tailor our saving policy accordingly. Nor do we consistently do what we ought to do. Suppose that future generations could be relied upon to try to follow UU. Would such angelic altruism, but without an archangel's foresight, trap us in an indefinitely postponed splurge?

Even then, zero time preference could not push us into drastic and perpetual saving. In part, this is because here again any material gains to later generations from Stakhanovism would be outweighed by cultural and ecological impoverishment. But there is another reason. If UU told us to defer consumption at time t_1, and we had exactly the same reason to defer consumption at t_2, $t_3 \dots t_\infty$, then UU would tell us to defer consumption at each of these following points.[21] We saw earlier that the action that is objectively right—what one

[19] Philibert, 1999.

[20] Cf. Tomasik, 2015, pp. 144–6. One worry that this raises is that equal consideration of wild animals' interests could become very demanding indeed. This is a problem for utilitarianism in general, however, and not the result of rejecting discounting. See Hills, 2010.

[21] Lagerspetz, 1999, p. 156.

would do if one had perfect information—must be the same for all agents in relevantly similar situations, because moral judgments supervene on non-moral facts. The same is true for judgments about what is subjectively right—what, given our available information, we can justifiably do.[22] When in relevantly similar positions, with access to the same facts, agents subjectively ought to do the same thing.

If each generation did as UU instructed, then deferring at t_1 would have very *low* expected utility. There is no expected utility in saving money that we know will never be spent. But then, it could not be right on UU to defer. Stated more formally:

(1) Actions should maximize expected utility.
(2) If a generation has the same reasons for and against deferring consumption at $t_1, t_2, t_3 \ldots t_\infty$, then if it should defer it at t_1, subsequent generations should defer it at $t_2, t_3 \ldots t_\infty$.
(3) The reasons will be the same at $t_1, t_2, t_3 \ldots t_\infty$.
(4) Each generation always does as it should.
(5) ∴ If a generation should defer consumption at t_1, then subsequent generations both should and will defer it at $t_1, t_2, t_3 \ldots t_\infty$.
(6) If the agents will defer consumption at $t_2, t_3 \ldots t_\infty$ then deferring it at t_1 fails to maximize expected utility.
(7) ∴ It is false that the agents should defer consumption at t_1 (by 1).

If we deferred consumption in the belief that it would have greater expected utility later, then we would ignore the future options with which our descendants would be confronted. Deferring consumption might increase expected utility if we were to save this year and dis-save the next. But it would not do so as part of an endless series of deferrals—and if we could foresee this would happen, we should not start down that path. Indeed, it would be better to dis-save at *any* point than never to consume at all.[23]

UU is not indirectly collectively self-defeating. The real problem is that it gives us only the vaguest guidance about what to do. Since the world is unlikely to end tomorrow, we can safely assume that the optimum savings rate will be greater than zero. Since Stakhanovism would do too much, we also know that it will be moderate. Yet beyond that, a wide range of rates seem equally plausible candidates. Perhaps we should conclude with Narveson that "what we owe to future generations is neither Everything nor Nothing, but merely Something."[24] That is not much in the way of a guide.

[22] For the objective-subjective distinction, see Parfit, 1987, 25.
[23] Arntzenius et al., 2004.
[24] Narveson, 1978, p. 60.

10.3. Discounting as a Utilitarian Decision-Making Procedure

Does this vindicate discounting after all? To avoid jam tomorrow, we must choose *some* reasonable balance of saving and consumption. Otherwise we would end up like Buridan's ass, who starved between two haystacks. Utilitarians have long recognized that maximizing utility can be hard to do. Agents can often do best by relying on "rules of thumb."[25] It might seem that the right savings rate will be more or less what discounting as currently practiced already gives us. Given a number of possibilities that seem equally likely to maximize utility in the long term, one might argue, we should choose the one that reflects the market interest rate. Otherwise we shall make inefficient investments that leave future generations worse off. Market rates do not tell us how much to save, so a common argument goes, but they do help us to allocate it where it has the highest rate of return.[26]

Such a defense of discounting, it is important to note, would already be quite different from an appeal to PIPS. Nevertheless, it is worth considering. This rationale holds that the market rate of return reflects the opportunity cost of investments. It would be foolish to invest in ambitious greenhouse gas mitigation if the same money, invested over time, would grow enough to compensate for the loss and still have resources left over. Thus Martin Weitzman argues that while climate change may cause the seas to rise, "a very modest savings program started now would accumulate enough bricks and metal and so forth that we could easily afford to build the walls and pumps and everything else we would need *if the underlying trend of the real rate of interest remains about the same.*"[27]

One well-known problem with this argument leaps out from Weitzman's caveat at the end. We cannot assume that the rate of return will remain the same. Should the earth suffer runaway climate change or some other catastrophe, long-term growth could stall, or the damages could be of a kind for which alternative consumption opportunities simply could not compensate.[28] Weitzman recognizes this problem, but suggests planning with our best estimate of what the rate of return will be, with the plan to update our estimates in light of new information.[29] Such a strategy would make sense if our emissions were like taps in a bathtub, where the water could be shut off when it neared the edge. Climate change, unfortunately, depends

[25] Bales, 1971.

[26] Sunstein, 2007, p. 12; Posner and Weisbach, 2010, pp. 145–68; for discussion, see Arrow et al., 1996, pp. 130–3; Meyer and Roser, 2012.

[27] Weitzman, 1999, p. 24. Emphasis in original.

[28] Dietz et al., 2007, p. 137.

[29] Weitzman, 1999, p. 27.

on stocks as well as flows, and cannot be just switched off, as Weitzman himself later pointed out.[30]

A second problem is that though economic growth may allow us to build walls and pumps, it will do little to help animals, who will suffer the lion's share of the damages.[31] Despite roots in the utilitarian tradition, economic cost–benefit analyses typically give little attention to animal suffering, often valuing it by human beings' willingness to pay to prevent it. This is rather like valuing the suffering of slaves through a survey of their owners.[32] If global warming forces starving polar bears to eat their own young,[33] this is bad in itself and not only because it upsets newspaper readers. Economic growth for humans will not shield most of these non-humans from damages, and may even make their situation worse.

Discounting as usually practiced, then, does not maximize long-run expected utility. Since it accords next to no weight to costs and benefits that arrive more than a century or two in the future, it would be surprising if it did.[34] At a typical discount rate, Graciela Chichilnisky observes, "[a] simple computation shows that if one tried to decide how much it is worth investing in preventing the destruction of the earth 200 years from now on the basis of measuring the value of foregone output, the answer would be no more than one is willing to invest in an apartment."[35] How could a decision-making procedure that encourages us to disregard such disasters be justified on utilitarian grounds?

10.4. "Just Keep Discounting, But..."

Discounting might work as a utilitarian decision-making procedure if we were sure that the world economy would continue to grow, and that growth could reliably compensate for ecological damage. Unfortunately, it would be Panglossian to rely on either being the case. Once per capita incomes exceed about $20,000 per year, additional gains in consumption add little to societies' happiness.[36] Runaway climate change, on the other hand—or a comparable disaster such as global thermonuclear war—could lead to a collapse in worldwide populations and welfare. From a utilitarian perspective, it is far more important to minimize the risk of such catastrophes than to increase consumption. The

[30] Stern, 2009, pp. 91–2; Caney, 2009, p. 173; Wagner and Zeckhauser, 2012, p. 511; Weitzman, 2011, p. 290.

[31] Hsiung and Sunstein, 2006; Cripps, 2013, p. 86. [32] Jamieson, 2014, p. 134.

[33] *Spiegel Online*, 28 November 2009. [34] Davidson, 2015, p. 402.

[35] Quoted in Gardiner, 2011, p. 268. On my own calculation, at a rate of 5 percent, it would take closer to 400 years, but the conclusion is absurd either way.

[36] Layard, 2005, pp. 32–3; Ng, 2011, p. 110; cf. Cowen, 2007, pp. 23–6.

main ways we can increase aggregate utility are to ensure that sentient beings continue to exist in the future with a sufficient standard of living, and, provided these conditions are met, that there should be more rather than fewer of them.[37]

We might try to reform the practice of discounting so as to take this into account. Geir Asheim has recently proposed that we discount costs and benefits to future generations only when they are better off than we are.[38] In many cases, however, we have some reason to believe a policy might be disastrous, but only the faintest inkling of the likelihood, the duration, or the extent of the damages.[39] Our inability to foresee how long life would otherwise survive surely would complicate any such calculations. Alternatively, we might assign infinite weight to damages that pose an existential threat, so that discounting cannot make them disappear. In effect, this may be Martin Weitzman's approach.[40] A third option would be to carry on discounting as usual, subject to the constraint of avoiding the risk of human extinction or humanity's permanent impoverishment.[41]

The last approach may be the most promising of the three. It would amount to adopting Hans Jonas's imperative of responsibility, which holds "[t]hat there *ought* to be through all future time... a world fit for human habitation, and that it ought in all future time to be inhabited by a mankind worthy of the human name," and which forbids us from making "the existence or the essence of man as a whole... a stake in the hazards of action."[42] Though Jonas based his imperative on a controversial non-consequentialist argument, it can be defended on thoroughly utilitarian grounds.[43] Giving due weight to non-human interests would remain a challenge, for it does not seem that we should, or even in practice could, give the same priority to preventing the extinction or impoverishment of other species.[44] Nevertheless, policies that minimize human extinction risk are likely to benefit animals as well. To borrow a phrase from Weitzman,[45] we would just keep discounting, but with a rule against any gambles with the future of the planet.

[37] Ng, 2011; Bostrom, 2013. It may be objected that the best way to maximize total utility could be to create a vast population with consumption spread very thinly (Parfit, 1986). But utilitarians can plausibly hold that a world in which everyone enjoys a satisfactory standard of living would be *lexically* superior. Such a view entails its own complications—to escape the "mere addition paradox" we must accept that value can be intransitive—but I have argued elsewhere that this is easier to swallow than its alternatives (Rendall, 2015).

[38] Asheim, 2012; Dietz and Asheim, 2012; Zuber and Asheim, 2012.

[39] Birnbacher, 1988, p. 157; Farber and Hemmersbaugh, 1993, pp. 295–6; Roser, 2009, p. 23.

[40] Weitzman, 2009. Antony Millner suggested this point to me in conversation.

[41] Farber and Hemmersbaugh, 1993, pp. 295–6.

[42] Jonas, 1984, pp. 10, 37, emphasis in original.

[43] Birnbacher, 2001, pp. 376–81; Gesang, 2011, pp. 157–8; see also Birnbacher, 1988, pp. 202–7.

[44] For the majority of many species' members, as things stand at present, life may not even be worth living. Horta, 2015; Tomasik, 2015.

[45] Weitzman, 1999.

10.5. Conclusion

For half a century, the paradox of the indefinitely postponed splurge has been cited as grounds for discounting future costs and benefits. Unless we discounted, so the argument goes, the benefits from investment to posterity would be so great that utilitarianism would enjoin a punishing savings rate. What would happen if we actually tried to benefit posterity in this manner? It would be a disaster—as Soviet history shows. At least as soon as societies escape severe poverty, Stakhanovism does more harm than good. It privileges material production at the expense of nature and culture, and can only be sustained through political repression. Even supposing extreme saving followed by dis-saving *could* maximize total utility across generations, our inability to foresee the end of the world would mean we would never cash in our investment in time. That is the kernel of truth in PIPS.

The odd thing is that economists have concluded this means we must discount, lest the expected benefits of saving tyrannize the present. "[T]oo much weight given to generations far into the future turns out to be self-defeating," warns Koopmans. "It does nobody any good."[46] But if extreme saving does no good, how could utilitarianism demand it? The problem is a different one—we know that the optimal savings rate will be moderate, but not exactly what it is. That need not keep us awake at night. Leaving the savings rate to the market seems to work tolerably well. The danger is that discounting will lead us to risk catastrophe—and here generations of the far future deserve their full weight.

Acknowledgment

I am grateful to participants in the Seventh Braga Meeting on Ethics and Political Philosophy for comments on an earlier version.

References

Arntzenius, F., A. Elga, and J. Hawthorne. 2004. "Bayesianism, Infinite Decisions, and Binding." *Mind* 113: 251–83.

Arrow, K. J. et al. 1996. "Intertemporal Equity, Discounting, and Economic Efficiency." In *Climate Change 1995: Economic and Social Dimensions of Climate Change,* edited by J. P. Bruce, H. Lee, and E. F. Haites, 125–44. Cambridge: Cambridge University Press.

[46] Koopmans, 1967, p. 9.

Asheim, G. B. 2012. "Discounting While Treating Generations Equally." In *Climate Change and Common Sense: Essays in Honour of Tom Schelling*, edited by R. W. Hahn and A. Ulph, 131–46. Oxford: Oxford University Press.

Bales, R. E. 1971. "Act-Utilitarianism: Account of Right-Making Characteristics or Decision-making Procedure?" *American Philosophical Quarterly* 8: 257–65.

Birnbacher, D. 1988. *Verantwortung für zukünftige Generationen*. Stuttgart: Reclam.

Birnbacher, D. 2001. "Wärmetod oder nuklearer Winter? Die zeitgenössische Apokalyptik in ethischer Perspektive." In *Endzeitvorstellungen*, edited by B. Haupt, 359–83. Düsseldorf: Droste.

Bostrom, N. 2011. "Infinite Ethics." *Analysis and Metaphysics* 10: 9–59.

Bostrom, N. 2013. "Existential Risk Prevention As Global Priority." *Global Policy* 4: 15–31.

Caney, S. 2009. "Climate Change and the Future: Discounting for Time, Wealth, and Risk." *Journal of Social Philosophy* 40: 163–86.

Cowen, T. 2004. "Resolving the Repugnant Conclusion." In *The Repugnant Conclusion: Essays on Population Ethics*, edited by J. Ryberg and T. Tännsjö, 81–97. Dordrecht: Kluwer.

Cowen, T. 2007. "Caring About the Distant Future: Why it Matters and What it Means." *University of Chicago Law Review* 74: 5–40.

Cripps, E. 2013. *Climate Change and the Moral Agent: Individual Duties in an Interdependent World*. Oxford: Oxford University Press.

Davidson, M. D. 2015. "Climate Change and the Ethics of Discounting." *Wiley Interdisciplinary Reviews: Climate Change* 6: 401–12.

Dietz, S., and G. B. Asheim. 2012. "Climate Policy Under Sustainable Discounted Utilitarianism." *Journal of Environmental Economics and Management* 63: 321–35.

Dietz, S., C. Hope, N. Stern, and D. Zenghelis. 2007. "Reflections on the Stern Review (1): A Robust Case for Strong Action to Reduce the Risks of Climate Change." *World Economics* 8: 121–68.

Farber, D. A., and P. A. Hemmersbaugh. 1993. "The Shadow of the Future: Discount Rates, Later Generations, and the Environment." *Vanderbilt Law Review* 46: 267–304.

Gardiner, S. M. 2011. *A Perfect Moral Storm: The Ethical Tragedy of Climate Change*. New York: Oxford University Press.

Gesang, B. 2011. *Klimaethik*. Berlin: Suhrkamp.

Hare, R. M. 1961. *The Language of Morals*. Oxford: Clarendon Press.

Hare, R. M. 1981. *Moral Thinking: Its Levels, Method, and Point*. Oxford: Clarendon Press.

Heath, J. 2017. "Climate Ethics: Justifying a Positive Social Time Preference." *Journal of Moral Philosophy* 14: 435–62.

Hills, A. 2010. "Utilitarianism, Contractualism and Demandingness." *Philosophical Quarterly* 60: 225–42.

Horta, O. 2015. "The Problem of Evil in Nature: Evolutionary Bases of the Prevalence of Disvalue." *Relations: Beyond Anthropocentrism* 3: 17–32.

Hsiung, W., and C. R. Sunstein. 2006. "Climate Change and Animals." *University of Pennsylvania Law Review* 155: 1695–740.

Jamieson, D. 2014. *Reason in a Dark Time: Why the Struggle Against Climate Change Failed—and What It Means for Our Future*. New York: Oxford University Press.

Jonas, H. 1984. *The Imperative of Responsibility: In Search of an Ethics for the Technological Age*. Chicago: University of Chicago Press.

Koopmans, T. C. 1967. "Objectives, Constraints, and Outcomes in Optimal Growth Models." *Econometrica* 35: 1–15.

Lagerspetz, E. 1999. "Rationality and Politics in Long-Term Decisions." *Biodiversity and Conservation* 8: 149–64.

Landesman, C. 1995. "When to Terminate a Charitable Trust?" *Analysis* 55: 12–13.

Layard, R. 2005. *Happiness: Lessons from a New Science*. New York: Penguin.

Mendelsohn, R. 2008. "Is the *Stern Review* an Economic Analysis?" *Review of Environmental Economics and Policy* 2: 45–60.

Meyer, L. H., and D. Roser. 2012. "The Timing of Benefits of Climate Policies: Reconsidering the Opportunity Cost Argument." *Jahrbuch für Wissenschaft und Ethik* 16: 179–214.

Narveson, J. 1978. "Future People and Us." In *Obligations to Future Generations*, edited by R. I. Sikora and B. Barry, 38–60. Philadelphia: Temple University Press.

Ng, Y.-K. 2011. "Consumption Tradeoff Vs. Catastrophes Avoidance: Implications of Some Recent Results in Happiness Studies on the Economics of Climate Change." *Climatic Change* 105: 109–27.

Parfit, D. 1986. "Overpopulation and the Quality of Life." In *Applied Ethics*, edited by P. Singer, 145–64. Oxford: Oxford University Press.

Parfit, D. 1987. *Reasons and Persons*. Oxford: Clarendon Press.

Parfit, D. 1995/2002. "Equality or Priority?" Reprinted in *The Ideal of Equality*, edited by M. Clayton and A. Williams, 81–125. Basingstoke: Palgrave Macmillan.

Philibert, C. 1999. "The Economics of Climate Change and the Theory of Discounting." *Energy Policy* 27: 913–27.

Posner, E. A., and D. Weisbach. 2010. *Climate Change Justice*. Princeton, NJ: Princeton University Press.

Rendall, M. 2015. "Mere Addition and the Separateness of Persons." *Journal of Philosophy* 112: 442–55.

Revesz, R. L. 1999. "Environmental Regulation, Cost–Benefit Analysis, and the Discounting of Human Lives." *Columbia Law Review* 99: 941–1017.

Roser, D. 2009. "The Discount Rate: A Small Number with a Big Impact." In *Applied Ethics: Life, Environment and Society*, 12–25. Sapporo: Center for Applied Ethics and Philosophy, Hokkaido University.

Stern, N. 2009. *A Blueprint for a Safer Planet: How to Manage Climate Change and Create a New Era of Progress and Prosperity*. London: The Bodley Head.

Sunstein, C. R. 2007. *Worst-Case Scenarios*. Cambridge, MA: Harvard University Press.

Tarsney, C. 2017. "Does a Discount Rate Measure the Costs of Climate Change?" *Economics & Philosophy* 33: 337–65.

Tomasik, B. 2015. "The Importance of Wild-Animal Suffering." *Relations: Beyond Anthropocentrism* 3: 133–52.

Von Laue, T. H. 1981. "Stalin Among the Moral and Political Imperatives, or How to Judge Stalin?" *Soviet Union/Union Sovietique* 8: 1–17.

Wagner, G., and R. J. Zeckhauser. 2012. "Climate Policy: Hard Problem, Soft Thinking." *Climatic Change* 110: 507–21.

Weitzman, M. L. 1999. "Just Keep Discounting, But..." In *Discounting and Intergenerational Equity*, edited by P. R. Portney and J. P. Weyant, 23–9. Washington, DC: Resources for the Future.
Weitzman, M. L. 2009. "On Modelling and Interpreting the Economics of Catastrophic Climate Change." *Review of Economics and Statistics* 91: 1–19.
Weitzman, M. L. 2011. "Fat-tailed Uncertainty in the Economics of Catastrophic Climate Change." *Review of Environmental Economics and Policy* 5: 275–92.
Zuber, S. and G. B. Asheim. 2012. "Justifying Social Discounting: The Rank-Discounted Utilitarian Approach." *Journal of Economic Theory* 147: 1572–601.

11

The Controllability Precautionary Principle: Justification of a Climate Policy Goal Under Uncertainty

Eugen Pissarskoi

11.1. Introduction

In order to justify a climate policy goal, foreknowledge about the consequences of climate change is required. However, we can forecast merely possible impacts of climate change. For we do not know objective probabilities for relevant parameters: climate sensitivity (CS) is one of them.[1] Thus, we face an epistemic situation which Knight (1921) called "situation under uncertainty": we know (some or all) possible consequences of the available action strategies but do not possess knowledge about their objective

[1] Indeed, climate scientists calculate probability density functions (PDFs) for the values of CS from which the IPCC deduces that the CS is "likely between 1.5°C and 4.5°C" (Stocker et al., 2013, p. 924). More recent estimations of the magnitude of climate sensitivity by the means of global climate models result in similarly broad ranges: Cox et al. (2018) report a likely range of CS between 2.2°C and 3.4°C; Brown and Caldeira's (2017) estimations result in a 25–75 percent interval of 3.0°C to 4.2°C.

However, several authors have argued that these probability distributions calculated by the means of climate modeling should not be interpreted as providing information about objective probabilities (e.g. Stainforth et al., 2005; Betz, 2007; Stainforth et al., 2007; Frigg et al., 2013). Rather, they provide knowledge about possible ranges of CS without additional information on the distribution of probabilities within these ranges.

Friedrich et al. (2016) derive CS magnitude from a reconstruction of historical data for global mean temperature, radiative forcing, and further parameters from paleorecords. They report a range for CS between 4.3°C and 5.4°C. These—comparatively high values for CS gained by an additional scientific method (inductive reasoning from historical observations)—provide a reason for being cautious in narrowing the range of CS magnitude and in taking the probability distributions from climate modeling at face value.

probabilities. By which normative principle can we reasonably justify a climate policy target, e.g. a target for the greenhouse gases (GHG) concentration, in that epistemic situation?

One reply to this question, called "probabilism" (Jeffrey, 1956), aims at converting uncertain consequences into risky ones by assigning subjective probabilities to them and applying methods of Bayesian learning—a procedure that allows using probabilistic decision theory (e.g. expected utility theory) and which is broadly accepted within economics (Heal and Millner, 2014). Subjective probabilities can be derived from degrees of beliefs about possible consequences. Savage (1954) showed that degrees of belief, which comply with certain rationality requirements (Savage's axioms), represent probabilities.

However, in this chapter I shall lay aside probabilistic approaches toward decision justification under uncertainty. I take Ellsberg's experiments to be a serious motivation for a quest for non-probabilistic principles for decisions under uncertainty. Ellsberg (1961) provided evidence that in certain decision situations under uncertainty people tend to violate a rationality axiom, the sure-thing-principle, that is required in order to interpret degrees of belief as probabilities. Thus, it is not clear whether it is always rational to comply with the rationality requirements of the expected utility theory under uncertainty.

Several authors have bitten the bullet of Knightian uncertainty with regard to climate change and have justified a climate policy goal by a "possibilistic approach," i.e. without reference to probabilities but merely on the basis of knowledge about possible outcomes (e.g. McKinnon, 2009; Shue, 2010; Hartzell-Nichols, 2012; Steel, 2013; Pissarskoi, 2014). They have suggested different versions of the precautionary principle (PP) therefore. Although they differ in the details of an explication of the PP, the main idea of their argumentation rests on a comparison of the worst possible consequences of the relevant action options.

Being sympathetic to the reasoning on the basis of a PP for a climate change policy, I disagree with the arguments brought forward so far. In the first step, I shall present two versions of the precautionary principle and argue that at least one of them is morally plausible (Section 11.2). Then, I shall reconstruct an argument for a climate mitigation policy as it is presented in the literature which takes the morally plausible version of the PP as a premise (Section 11.3). I shall object to this argument that it underestimates the possible range of CS. If we fully take into account the possible magnitude of CS, the sole policy option by which we will avoid catastrophic impacts is a reduction of GHG concentrations to the pre-industrial level (280 ppm CO_2e) as quickly as possible, or so I shall argue (Section 11.4). However, the PPs suggested so far do not justify a choice between the climate policy goal of 280 ppm and a moderate climate policy goal (Section 11.5). Finally, I shall suggest a further version of the precautionary principle, the controllability PP, which I think better suits the characterized epistemic situation, and I shall discuss its moral plausibility (Section 11.6).

11.2. Two Versions of the Precautionary Principle: SPHL-PP and RCPP

A prominent type of risk-averse principles for decisions under Knightian uncertainty are PPs.[2] Their common idea states that under Knightian uncertainty there is a normative requirement to choose cautious actions if the outcomes of the available actions can severely harm others (c.f. Wingspread Conference Participants 1998). More precisely, PPs are conditional statements with a certain structure (Sandin, 1999, 2004; Gardiner, 2006; Steel, 2015).

If an agent is under the following conditions:
Epistemic: she foresees merely possible outcomes of the available options among actions, and
Evaluative: some outcomes possibly lead to severe harm,
then the agent is under a certain normative requirement[3] to choose an action that avoids possible outcomes that are harmful.

This general structure allows explication of a broad variety of PPs, depending on the specification of (i) the epistemic situation, (ii) possible consequences and their evaluation, (iii) the normative requirement, and (iv) the decision rule (rule specifying which action option to choose).

With regard to climate change, several authors have argued for an ambitious climate mitigation policy on the basis of what I call the "serious-possibility-of-high-losses" precautionary principle (SPHL-PP) (McKinnon, 2009; Shue, 2010; Hartzell-Nichols, 2012; Steel, 2013). I shall reconstruct it according to the structure above:

1. Epistemic conditions:
   (a) decision makers are in a situation of Knightian uncertainty and
   (b) decision makers take into account only outcomes which are seriously possible—Shue calls this condition the "anti-paranoia requirement" (Shue, 2010, p. 149).

[2] Other types of decision principles under uncertainty would be risk-prone decision principles. They explicate circumstances under which it is normatively justified to choose a risky action option (e.g. based on a formal decision rule like the maximax).
[3] E.g. she is morally obliged or is permitted.

2. Evaluative conditions:
   (a) it is possible that some available acts will lead to severe harm;
   (b) there are alternative options among actions whose worst possible outcomes avoid severe harm and do not impose other morally relevant losses.
3. Action requirement: under these epistemic and evaluative conditions decision makers ought to avoid action options which can lead to severe harm.

The second epistemic condition (1b), the anti-paranoia requirement, determines which kind of possible consequences should be taken into consideration within the decision-making process. The reason for this condition is to avoid taking irrelevant possible outcomes, e.g. merely logically possible consequences, into account. Shue and Steel suggest similar explications for the anti-paranoia requirement, i.e. for what they consider as "seriously possible" outcomes. Shue's PP takes into account possible consequences whose

(i) mechanism by which the impacts would occur is "well understood" and
(ii) "the conditions for the functioning of the mechanism are accumulating" (Shue, 2015, p. 87).

Steel considers possible consequences for which what he calls a "scientifically plausible mechanism" (Steel, 2013, p. 329) is known, meaning that

(i) the possible outcome is "grounded in scientific knowledge" and
(ii) "scientific evidence exists for its actual occurrence" (Steel, 2013, p. 329).

Besides the term "serious possibilities," the conditions of the PP contain further thick moral concepts ("well-understood mechanism," "relevant losses," "severe harm") which need an explication. But I do not think that this poses an insurmountable difficulty for the principle—it seems to me to be a rather technical task to plausibly sharpen the thick terms.[4]

Is it morally reasonable to accept the SPHL-PP? The moral intuition behind this version of the PP is a transfer of the moral requirement "avoid serious harm" which is applicable to a deterministic epistemic situation to a situation

[4] Peterson (2006) argues that the PP is inconsistent as a decision rule. I doubt that Peterson's criticism provides an objection to the PP versions discussed here, for several assumptions of Peterson's impossibility theorem do not hold in the decision situation discussed in this chapter. First, Peterson considers decision situations in which the outcomes can be ordinally ranked according to their probabilities. However, here we are focusing on a decision situation in which the outcomes' probabilities cannot be ordinally ranked. The outcomes lie within a certain epistemic range (they are seriously possible), but no further (objective) probabilities or likelihoods, even qualitative ones, can be assigned to them. Secondly, as Steel (2015, p. 41) has argued, the condition "Archimedes" in Peterson's impossibility theorem is not plausible for decision problems under uncertainty, especially for cases in which outcomes' probabilities cannot be ranked cardinally.

of uncertainty. The SPHL-PP makes explicit which conditions have to obtain in order for a moral agent to be under the moral requirement to avoid possible harm to others: it is seriously possible that some of your acts will substantially harm other morally relevant beings; there are other action options available by which you would avoid harming others without imposing significant hardships to you. Under these conditions there can be no excuses not to choose a cautious option among actions. This is the normative idea behind the SPHL-PP.

However, I doubt that under the stated conditions we are always morally required to choose the cautious action option. For the SPHL-PP justifies the decision on the basis of the worst possible consequences only. Suppose the following case. You are in a decision situation between two action options, A and B. About the consequences of the options you have only possibilistic foreknowledge. If you choose A, in the worst possible case some massive losses to future generations will occur. But in the best possible case these generations will gain huge benefits (e.g. a distributively just society at a high level of quality of life will be established). If you choose option B, then in the worst possible case you will avoid the massive losses to future generations without imposing high mitigation costs for you. But in the best possible case the future generations will still live in a similarly unjust global society as the current one.

In this decision situation, the epistemic and the evaluating conditions of the SPHL-PP are fulfilled. However, I doubt that we are morally required to choose the precautionary option B. For by choosing it we preclude future generations from the possibility to attain presumably the highest social good: a just society. I do not, however, claim that option A ought to be chosen. In order to come to a morally justified decision we would have to balance between the possibility of high losses and the possibility of high gains. Therefore, we would need more information about the case than I provided here. For instance, we would need to know by which (social, political, economic, natural) mechanisms a just society will be established if we choose A, how well understood these mechanisms are, whether they are controllable and by whom they are controllable, etc.

Here I am stressing the point that there do exist situations in which the epistemic and evaluating conditions of the SPHL-PP are fulfilled, and we are not morally required to choose action options that avoid the possibility of severe harm. The possible benefits from the available options (i.e. the value of the best possible consequences) might possess moral weight and should be accounted for within the conditions of the PP.

The version of the PP that Stephen Gardiner (2006) suggested, the Rawlsian core precautionary principle (RCPP), accounts for this point. Gardiner refers to Rawls's explication of the conditions under which the application of the maximin decision rule is justified (Rawls, 1971, pp. 133f.). The RCPP differs from the SPHL-PP mainly by a further evaluative condition that assesses the best possible consequences:[5]

2. (c) The highest possible benefits from the available action options are not significant.[6]

Thus, while the SPHL-PP considers only the worst possible outcomes, the RCPP includes also a balancing of the negative with the positive consequences: it applies to decision situations in which some available action options will possibly violate moral rights and in which no acting option can bring about morally significant benefits. According to the RCPP, we are morally required to choose in such a situation the option among actions whose worst possible outcome would be the best compared to the worst possible outcomes of the other available action options.

Let us assume that it is reasonable to accept the RCPP and consider this version of the PP to be an example for a morally justified principle for decisions under Knightian uncertainty. Does it allow specifying which climate policy goal to choose?

11.3. Justification of a Climate Policy by the RCPP and the SPHL-PP

Since we can forecast merely possible outcomes of climate change, the condition (1a) of the PP is fulfilled. The RCPP as well as the SPHL-PP require that we evaluate the seriously possible outcomes of the available options (condition 1b).

If we mitigate modestly or do not mitigate at all, it is seriously possible that catastrophic impacts will occur in the worst case: Steel (2013, p. 329) quotes the IPCC reports showing that even an increase of the global surface temperature by 2°C can severely harm hundreds of millions of people. McKinnon (2009) and Shue (2010, 2015) argue that a modest mitigation strategy risks that large-scale environmental systems will be brought out of their current state: e.g. the thermohaline circulation will be disrupted; large ice sheets (West Antarctic, East Antarctic, and the Greenland Ice Sheet) melt down (Shue, 2015, p. 93) or methane clathrates in the oceans will be destabilized by the

[5] There are further differences between the RCPP and the SPHL-PP. However, since they do not matter for my objections to these versions of the PP, I mention them briefly: (i) the anti-paranoia condition in Gardiner's version is more vague. He indicates that the decision makers should consider only "credible threats" (Gardiner, 2006, p. 51); (ii) the action requirement of the RCPP is on the other hand more precise: the decision maker ought to choose an act according to the maximin decision rule (Gardiner, 2006, p. 47).

[6] "Decision-makers care relatively little for potential gains that might be made above the minimum that can be guaranteed by the maximin approach" (Gardiner, 2006, p. 47).

water temperature rise and release the stored methane (McKinnon, 2009, p. 188). By such a large-scale environmental change "the majority of life on Earth, perhaps including homo sapiens" could become extinct (McKinnon, 2009, p. 188). And, so the authors stress, these possible outcomes are not merely contrived in an armchair, but qualify as what they respectively understand as "seriously possible": we do scientifically understand the mechanisms by which the thermohaline circulation can be disrupted or large ice sheets melt (McNeal et al., 2011, pp. 666f.); and, secondly, evidence supporting the hypothesis that rising GHG concentrations are destabilizing large-scale environmental systems (e.g. with regard the deglaciation of West Antarctica) is accumulating (Shue, 2015, pp. 90ff.). Steel (2013, p. 329) argues that the correspondent mechanisms are scientifically plausible because the IPCC reports the possibility that a temperature rise above 2°C can lead to catastrophic impacts. Therefore, they meet the criteria for the seriousness of this possible scenario according to Steel. Thus, condition (2a) is fulfilled, too: it is seriously possible that in the worst case a modest mitigation strategy will lead to catastrophic consequences for morally relevant beings.

Then, the authors argue that there are alternative action options whose worst possible consequences avoid catastrophic impacts: an ambitious mitigation (e.g. limiting the increase in the surface temperature at 2°C above the pre-industrial level). According to these authors, there is broad economic evidence that the realization of such a mitigation target will not impose excessive burdens on the current generation (economic estimates vary between economic gains from mitigation efforts and costs of around 1 percent of global GDP annually, Shue, 2015, pp. 100ff.; Steel, 2013, pp. 331f.). Even in the worst possible case, ambitious mitigation will bring about relatively low economic costs: 4 percent of global GDP annually according to the highest estimates by the Stern Report (Stern, 2007). Although this amounts to considerable sums for climate policies, according to Steel and Shue, it is not seriously possible that additional investments amounting to 4 percent of global GDP will lead to severe catastrophes. For, although it is conceivable that such a mitigation strategy would end up in a disaster like a nuclear war (Manson, 2002), this possible scenario does not qualify as a serious possibility according to Steel's or Shue's explication of the term. There are neither well-understood mechanisms explaining how additional investments amounting to 4 percent of global GDP would lead to nuclear wars (or other catastrophic impacts) nor scientific evidence for it. Thus, with regard to the condition (2b) of the PP, the authors conclude that ambitious mitigation is an alternative option whose seriously possible worst outcomes avoid severe harm and do not impose any other morally relevant losses.

The condition (2c) of the RCPP is not addressed explicitly by the authors discussed here, so let us assume that they implicitly presuppose this condition to be satisfied, which means that the best possible consequences of the available climate change strategies are not important.

Given these conditions, the RCPP and the SPHL-PP morally require us to mitigate ambitiously.

However, I disagree with this argument. Although the SPHL-PP (including the condition 2c) and the RCPP are morally justified, they do not determine a choice among an ambitious and a modest climate policy. For the condition (2b) of these PPs is not fulfilled in the decision situation presented. In Section 11.4, I shall argue that it is seriously possible that mitigation strategies, which are ambitious enough to avoid catastrophic impacts from climate change, will impose catastrophic outcomes (from the transformation of economic and social systems).

11.4. Objections to the RCPP and SPHL-PP Arguments

CS represents how the climatic system reacts to an increase in radiative forcing (from doubling the GHG concentration from the pre-industrial level of 280 ppm CO_2e to 560 ppm). However, for all we know, its value lies somewhere within a broad range. Climate models provide evidence for a range between 1°C and double-digit values (e.g. 11°C resulted in some model runs, Stainforth et al., 2005).[7] Therefore, it is seriously possible that CS amounts to 11°C according to the explications suggested by Shue and Steel. First, the value is a result of scientific research. Second, the main underlying causal mechanisms (between an increase in radiative forcing and the reaction of climate system, for which global surface temperature serves as an indicator) are well understood. Third, the temperature increases in surface temperature observed so far do not contradict the hypothesis that CS lies in the double-digit area.[8]

Since it is seriously possible that CS amounts to a double-digit value, I doubt that the PPs presented above, the RCPP and the SPHL-PP, help us to discriminate between alternative policies and single out one policy that is justified. I will now explain why.

For the sake of simplicity, let us focus on a decision situation between two action options: a modest mitigation one as suggested by the *Stern Review* (concentration target of 550 ppm CO_2e by 2050, Stern, 2007) and an ambitious target: reduction of the greenhouse gas concentrations to the pre-industrial

[7] The IPCC presents probability distributions of climate sensitivity and does not report the possibility range of modeling results. However, the graphical representation of the probability density functions indicates that several models verify the possibility of CS equalling 10°C (Stocker et al., 2013, p. 925, figure 10.20b).

[8] The CS results from models that are calibrated on the basis of available evidence and thus are consistent with the observations so far.

level (280 ppm CO_2e) by 2050 (it will become clear in a moment why I am suggesting the pre-industrial concentration level as a policy target).

In order to characterize possible outcomes of these two action options I shall refer to different types of possibility statements distinguished by Betz (2010). First, some possibility statements are verified. That means it has been shown that a possibility statement is consistent with the available relevant background knowledge. Serious possibilities according to Shue's and Steel's explication are a subset of verified possibilities.[9] Second, some possibility statements are falsified (i.e. an impossibility statement has been verified). That is the case if and only if it has been shown that a possibility statement contradicts some truths. Third, there is a set of articulated possibility statements which are neither verified nor falsified. Betz calls them possibilistic hypotheses (Betz, 2010, p. 93; 2015, p. 195).

Let us consider the 550 ppm target first. Its normatively relevant consequences depend on the reaction of natural systems and on the changes in social and economic systems caused thereby. In the best possible case, CS is low, according to climate modeling results about 1°C (Stocker et al., 2013, p. 925, figure 10.20b). Stabilization of GHG concentrations at 550 ppm is not expected to impose severe harms in this case. Economists provide evidence that this stabilization goal can be achieved nearly without additional costs (Edenhofer et al., 2014, p. 449) or even bring economic benefits (e.g. Barker and Scrieciu, 2010).

On the other extreme, it is verified by climatic modeling that CS can amount to 11°C. This implies that the 550 ppm target will lead to a temperature increase of nearly 11°C in the long run. It is possible that such a temperature rise will exceed several critical thresholds in natural systems, the so-called "tipping points" (e.g. melting of polar ice sheets, collapse of Atlantic meridional overturning circulation (AMOC), release of carbon from permafrost or of methane from clathrates (Stocker et al., 2013, p. 1115)). There is evidence from AMOC modeling that it is possible that the circulation will shut down in the long run if surface temperature rise amounts to double-digit values (Stocker et al., 2013, p. 1094, figure 12.35). We also have historical evidence for exceeding some tipping points (e.g. melting of ice sheets, thawing of permafrost, and release of methane at a temperature change of around 8°C, c.f. McNeall et al., 2011). Thus, it is a verified possibility that some tipping points in natural systems will be exceeded in the worst possible case.

How will these changes in natural systems impact social systems? To my knowledge, there are no scientific assessments of social and economic impacts for scenarios with CS in a double-digit range. However, we know that even

[9] Note that Betz (2015, p. 195) suggests a different explication of the term "serious possibilities."

temperature increases of 4–6°C can bring about social consequences which we characterize as catastrophic: due to famines, migration, conflicts over natural resources, etc., a significant part of humanity will severely suffer. Therefore, it is a verified possibility that a 550 ppm target will bring about catastrophic impacts in the case that CS amounts to 11°C.

If we do not restrict ourselves to the verified possibilities but take into account possibilistic hypotheses, too, we can articulate the possibility of humanity's extinction by 2100 as the worst case. A temperature increase of this magnitude would cause huge migration waves, e.g. in East Asia (Bangladesh) or in the Middle East. These migration waves would destabilize states in these regions, causing military conflicts which could involve also powers possessing nuclear weapons. It is possible that such a military conflict would result in humanity's mass extinction. I do not know any truths contradicting this possibilistic statement. Therefore, it is a neither falsified nor verified possibilistic hypothesis.

Here lies the reason why I disagree with Shue's and Steel's judgments about whether condition (2b) is fulfilled. According to this condition, there should be a policy option among the available options for action whose realization would not impose moral harms. Steel and Shue claim that this holds for modest mitigation. However, the evidence from the economic literature they provide merely demonstrates that a 500–50 ppm strategy is affordable at relatively low costs. However, this mitigation goal does not suffice for the avoidance of catastrophic impacts even according to Shue's and Steel's understanding of "serious possibilities." We thus need to find a mitigation strategy which guarantees that seriously possible catastrophic impacts from climate change will be avoided.

High values of CS mean that natural systems very sensitively react to small perturbations in radiative forcing. The latest (2016) GHG concentration of 489 ppm CO_2e induces a radiative forcing of 3 W/m^2 (Butler and Montzka 2017). If CS amounts to a double-digit range, today's concentration already implies a long-term increase in global surface temperature of about 6–8°C. This would change global environmental systems substantially, leading to catastrophic harms for human beings and other species, even humanity's extinction is possible. I do not see any other mitigation goal than reduction of GHG concentrations to the pre-industrial level that can ensure that severe harm from climatic change will be avoided.

Let us now turn to the question of what are the best and the worst possible consequences from the climate policy target of 280 ppm CO_2e. Consider the worst case first. We deterministically know that the anthropogenic climate change impacts will be avoided in that case. However, we would have to substantially change our social and economic systems in order to realize it: to fully decarbonize energy production, change agricultural and livestock farming in order not to emit additional GHGs, and to implement means for removal of GHGs from the atmosphere. I do not know of any scientific forecasts of such an ambitious scenario. The most stringent mitigation scenarios the IPCC reports aim at 430–80 ppm CO_2e (Edenhofer et al., 2014, pp. 449f.). There is evidence from economic modeling that the costs of the 430–80 ppm target can rise exponentially compared to the 500–50 ppm target (Kriegler et al., 2014).

However, the threats of an ambitious mitigation strategy result not only from economic costs but also from side effects of the deployment of negative emissions technologies (NETs) (Fuss et al. 2014). Technologies such as bioenergy with carbon capture and storage (BECCS) or afforestation (AR) require agricultural area and water in a considerable amount. Smith et al. (2016) calculate the resource implications of the NETs under the assumption that they will be implemented in an amount that is compatible with the 430–80 ppm target. According to Smith et al. (2016, p. 46), a deployment of BECCS in that amount would require an agricultural area of 380–700 Mha which corresponds to 7–14 percent of the global agricultural area (or 25–46 percent of the global croplands) and an additional use of around 3 percent of human-appropriated fresh water. Similar numbers hold for afforestation. Other NETs (direct air capture (DAC), enhanced weathering of minerals (EW)) need less land area and water but have much higher energy requirements (DAC) or much smaller carbon removal potential (EW) (Smith et al., 2016, table 1). These calculations indicate that a deployment of NETs in an amount that is compatible with the 430–80 ppm target can lead to grave food and freshwater shortages.

This possibility counts as a serious one according to Shue's and Steel's classification. Accordingly, it follows that it is seriously possible that the 280 ppm target will produce extensive global food insecurities. And if we take into account that the BECCS deployment compatible with the 430 ppm target might require 25–46 percent of the arable land and permanent crop area (Smith et al., 2016, p. 46) it is not implausible to conceive that the attainment of the 280 ppm target can cause global social tensions finally ending up in military conflicts between powers with nuclear weapons. It is neither a verified nor a falsified possible statement that such conflicts will result in humanity's extinction in the worst possible case.

On the other hand, a more just society will be realized in the best possible case if we pursue the 280 ppm target. Advocates of the so-called "degrowth" strategies describe such a utopia (e.g. Jackson, 2009; D'Alisa et al., 2014): in order to reduce the GHG emissions and concentrations economic activity has to be downscaled. Although economic activity would shrink, we would reduce the average working hours and create new jobs within service sectors (e.g. the care economy), and people would have less income per capita but more leisure, which they would use for social and political activities thereby establishing solidarity and a just society. Similarly, it is neither a verified nor a falsified possibility that the social visions described by the "degrowth" advocates will be realized.

In summary, here is what we know about the possible outcomes of the two climate policy goals, 550 ppm and 280 ppm.

It is a verified possibilistic hypothesis that the realization of the 550 ppm target can lead to severe harm for a great amount of people resulting from environmental change. In the worst case this target might induce humanity's extinction. The latter statement expresses a possibilistic hypotheses: we do not know that it is consistent with our background knowledge; but it also does not contradict it. In the best case a stabilization of GHG concentrations at 550 ppm might bring about gains in economic wealth.

With regard to the 280 ppm target we know deterministically that it will avoid harm from climatic change. But its realization has its costs: it is a verified possibilistic hypothesis that it will induce severe harm. In the worst case, social conflicts can escalate up to humanity's extinction. We cannot falsify this possibility hypothesis on the basis of our background knowledge. In the best case, a radical transformation of global economies (compatible with the realization of the 280 ppm target) might create a more just social order than the current global society.

Now we can see more clearly why the SPHL-PP and the RCPP do not allow discrimination between climate policy options. First, these principles are epistemically too demanding. They ask us to compare the available options among actions on the basis of the best and the worst possible consequences. However, with regard to climate change impacts, we can forecast the possible consequences merely in coarse terms which do not allow differentiating among the options. For all we know, in the worst possible case the considered action options are on a par. And it does not matter which set of possibilities we consider: if we stick to what Shue and Steel explicate as serious possibilities, then the worst possible consequences of both climate strategies will be catastrophic impacts for human beings due to food shortages, water insecurities, etc. If we also take possibilistic hypotheses into account, then the worst possible consequence from both strategies is humanity's extinction. We cannot state the number of the affected persons or the severity of the shortages more precisely.

Second, both versions of the PP are restricted to a decision situation in which possible damages and benefits are distributed among options in a certain, morally benign, manner: some options possibly lead to grave losses and can bring about only tiny benefits; other options avoid grave losses in the worst possible case without imposing high implementation costs. It is actually this distribution of the positive and the negative possible outcomes that confers moral plausibility on the RCPP and SPHL-PP. Unfortunately, the decision situation with regard to climate change goals is more intricate: all available options can bring about catastrophic harms as well as significant benefits. So, how should we justify a decision in this intricate epistemic and evaluative situation?

11.5. Reasoning Within the Epistemic Situation

The inconvenience of our epistemic situation lies in the fact that we cannot order the possible consequences of the available options among actions with regard to their betterness. If the worst case comes true, it is possible that the outcome of the 280 ppm strategy will be better than the outcome of the 550 ppm strategy. But it is equally possible that the worst possible consequences of the ambitious climate mitigation strategy will be worse than the worst possible outcome of the 550 ppm action strategy.

That means that the decision between the action options has to be justified on the grounds of a decision principle that does not require information about the value of the consequences of the climate policy options. However, if there is to be a cognizable reason in favor of one of the action options, then there must be a cognizable difference between the two options. What could a relevant difference be?

One respect in which the action options differ is with regard to their political feasibility. Intuitively, the more ambitious goal (280 ppm) seems to be more difficult to implement politically than the moderate one. Thus, it might be suggested that in our epistemic situation this difference provides a sufficient reason in favor of the 550 ppm target.

However, this pragmatic difference gains normative weight only at the stage of political decision-making and after all other normative reasons have been assessed. For political feasibility can be changed to a certain degree by normative reasoning and public argumentation. As long as the available options are feasible under certain circumstances we need to know how the options are normatively weighted, i.e. which of them ought to be realized. On the basis of this knowledge we can then assess the trade-off between their feasibility and their ideal normative requirement in the next step.

I am interested in the ideal normative justification of the two options. Do we find any reasons that would justify the choice between them? And if not, does it then follow that pragmatic considerations should play a decisive role?

11.6. The Controllability Precautionary Principle

There is at least one further difference between the two options among actions. In realizing the ambitious goal of a concentration level of 280 ppm we definitely preclude a temperature change that would otherwise have been caused by greenhouse gases. We still can cause a lot of other dangerous interventions in natural systems, but at least about this one causal effect we know that it would disappear. This does not hold for the 550 ppm target. In choosing it we still nearly double the pre-industrial concentration level, and we know that this amount of GHGs can cause serious harm.

There is a reason to avoid an act that will trigger off a causal chain that can severely harm others. Therefore, the difference between the two concentration targets provides a reason in favor of the pre-industrial concentration level. Let us consider where its normative significance stems from.

If the GHG concentration will be reduced to the pre-industrial level, we will definitely avoid exceeding policy-relevant tipping points. Tipping points are thresholds in natural systems for which it holds that, if the threshold is crossed, a causal chain of reinforcing changes will be triggered that cannot be stopped or controlled and that leads to a new state of the natural system. Tipping points are policy-relevant if they have significant impacts for morally relevant beings (Lenton et al., 2008). Due to our uncertainty about the value of CS, it is possible that even small increases in GHG concentration exceed policy-relevant tipping points. Therefore, only by reducing the GHG concentration to the pre-industrial level do we exclude the possibility of crossing the critical thresholds.

Thus, the 280 ppm option allows avoiding triggering irreversible and non-controllable causal chains of changes in natural systems with catastrophic consequences, whereas the 550 ppm option does not guarantee this. And here lies the moral intuition in favor of the more ambitious climate target: under certain epistemic and evaluative conditions we ought not to trigger irreversible and non-controllable causal chains if their consequences can severely harm others.

Importantly, the ability to control processes that are induced by the chosen strategy and to influence their outcome is a capacity that gains normative significance in the situation of uncertainty. In a deterministic epistemic context, in which the available action options deterministically lead to certain outcomes, the ability to control the process does not play any role. This changes in a possibilistic epistemic situation. Since our action choices can lead to a broad range of possible outcomes, it becomes important to keep the ability to influence the process by which the outcomes are realized.[10]

[10] I believe the ability to control the process to have such normative importance that there are trade-offs between the value of the possible consequences and the ability to control the outcome process. We might find examples for decision situations under uncertainty in which it is reasonable not to choose an action option according to the maximin principle if there is an action option that is better controllable than the action option supported by the maximin. It remains a task for future work to discuss this trade-off between possible consequences and controllability.

Let us now consider what sufficient epistemic and evaluative conditions have to be fulfilled for the moral requirement to plausibly hold:

1. Epistemic conditions:
   (a) decision makers are in a situation of Knightian uncertainty and
   (b) anti-paranoia requirement: decision makers take into account only outcomes which are seriously possible.[11]
2. Evaluative conditions:
   (a) if the worst possible consequences of all available options will lead to severe harm,
   (b) if we cannot order the worst possible consequences of the available options with regard to their moral value, and
   (c) if we cannot order the best possible consequences of the available options with regard to their moral value.
3. Process conditions:
   (a) if it is possible that some available options cross a tipping point, that is, trigger off a non-controllable causal chain leading to an irreversible state with severely harmful consequences,
   (b) if there is at least one available option that avoids crossing any tipping points, that is, triggering off a non-controllable causal chain leading to an irreversible state with severely harmful consequences, and
   (c) if the decision maker is able to realize the option that avoids all tipping points without violating any basic moral rights.
4. Action requirement:
   the decision maker ought to choose the action option that avoids crossing a tipping point.[12]

Let us call this version of the precautionary principle the "controllability PP" (CPP). The first group of conditions explicates the epistemic situation, the second describes the value of the possible consequences. The previous discussion has shown that the epistemic and evaluative conditions alone do not signify a morally relevant difference between the action options we are

[11] This condition is not committed to Shue's or Steel's explication of "seriously possible." Betz's (2015, p. 195) suggestion to consider all statements being consistent with the entire background knowledge as seriously possible is also an explication that can be applied here.

[12] That version of the PP is adapted to the situation in which there is one action option that fully avoids crossing a tipping point. Here is a proposal for a more general version of this PP, therefore the condition (3b) and the action requirement should be substituted by the following statements:

3b*: if the available options can be ordered with regard to the number of tipping points that will be possibly crossed in the course of their realization

4*: then the decision maker ought to choose the option which will trigger off the least number of tipping points.

The question whether this generalization is justifiable needs its own discussion.

interested in. We found that the relevant normative difference lies in a feature of the process of their realization. Pursuing the 550 ppm target, we may exceed tipping points and thus trigger off causal processes in natural systems which are not controllable by humans and which would lead into catastrophes. Pursuing the 280 ppm target on the other hand, we avoid triggering off uncontrollable processes in natural systems with dangerous consequences. Therefore, the general structure of the PP has been amended by a further set of conditions, the process conditions.[13]

By introducing process conditions into the PP, we can discriminate among pathways of the available policy strategies. Although all available climate change policies can lead to a catastrophic outcome, their pathways differ in the causal mechanisms which influence the outcomes: runaway natural processes and runaway social processes. Decision makers can then evaluate the pathways of the strategies and choose the option which is less dangerous because it is more easily controlled.

Let us now discuss whether the CPP justifies a choice between the 550 ppm and the 280 ppm targets. The realization of the more ambitious climate target avoids triggering tipping points in natural systems. However, its realization requires radical socio-economic transformations (in order to change people's behavior in such a way that they emit less GHGs) and/or implementation of NETs about which we know that they compete with food and water provision. Radical social reforms and BECCS could exceed tipping points in social systems with fatal consequences. Thus, one could argue, the CPP faces the same difficulty as the other versions of the PP: it does not allow differentiation between the action options discussed, for we cannot distinguish the realization of which option is more controllable.

I disagree with this objection, although I have to admit not being able to provide a full-fledged justification of the claim that CPP supports the 280 ppm target. For the question—which realization process is more controllable—is empirical and has to be discussed by the relevant empirical scientists more carefully than I can do it here. Therefore I shall sketch general reasons why I do not believe that it is seriously possible that tipping points will be exceeded by pursuing the 280 ppm target.

Negative consequences from climate change can have three sources: either they result directly from changes in natural systems (e.g. additional storms or floods), from social or economic changes induced by the changes in natural systems (e.g. rise in prices of agricultural products), or from people's own

[13] The process condition (3b) entails an evaluation of the realization processes of the action options. If the available action options differ significantly in the value of their realization, then this aspect should be accounted for within the decision principle (it would have to be amended by a principle weighting the ability to control a process against the possibility of violating moral rights).

reforms of social or economic systems in reaction (or pro-action) to the natural changes (e.g. taxation of GHG emissions). These sources are different in two respects.

First, they differ in how well we can control or influence these sources of negative impacts: impacts from changes in natural systems are under human control to a lesser degree than impacts from changes in social or economic systems. We can control the level of the GHG gases in the atmosphere, but not what these gases cause after they have been emitted. In contrast, our social and economic institutions are intended, designed, and established by us.

Second, there is a difference in how severe the consequences from the loss of control are: if we exceed tipping points in global natural systems, the resulting consequences will be fatal and endanger humanity as a whole without us having any possibility of influencing the causal chain. The only strategy in order to alleviate the impacts would be adaptation to the changed environmental conditions. If we lose control of some socio-economic reforms, there are no causal chains leading to devastating impacts for humanity that we could not influence. We have opportunities to intervene and change the social institutions in order to avoid catastrophic outcomes. It is true that we cannot steer the consequences of social reforms precisely. History provides enough examples of political intentions and reforms which ended in the opposite of what was originally intended (c.f. Parker (2013) for political disruptions in the seventeenth century or Figes (1997) describing the Russian revolutions at the beginning of the twentieth century). That is the reason why we cannot exclude that ambitious climate mitigation strategies will end up disastrously. Still, the outcomes of social and economic transformations become apparent in a shorter time frame than interventions with natural systems. Thus, we can rapidly react to undesired consequences by further reforms. Therefore, social systems remain controllable to a greater degree than the natural ones. If these claims are true, they provide a justification of the claim that global society ought to reduce GHG concentrations to the pre-industrial level of 280 ppm as quickly as possible.

11.7. Practical Implications

What follows in practice from the recommendation to pursue the 280 ppm target? Practical implications from my conclusion actually do not differ from what other authors recommend who justify a climate policy goal by a PP: humanity ought to start reducing its GHG emissions immediately in the biggest possible steps while taking care that an economic (or other social) catastrophe is not created (e.g. Steel, 2013, p. 330; Shue, 2015, p. 96). If we consider GHG taxes as the best economic instrument in order to reduce GHG emissions, then the taxes should be designed in a way that they lead to the highest possible GHG emissions reductions without jeopardizing economic stability. Economists provide evidence that a lot could be done without seriously endangering economic systems: it is possible to design economic instruments such that mitigation costs in an amount of 3–4 percent of global GDP will not destabilize economies.

That principle holds also for individual behavior. Each individual ought to reduce GHG emissions for which she is responsible individually (e.g. emissions from our nourishment preferences, holiday behavior, etc.) in the biggest possible steps taking into account her conception of a meaningful life. Similar considerations apply to the deployment of NETs: to the degree that they do not endanger food and water security (or other normatively significant values) they should be implemented. We cannot predict whether the mitigation efforts according to these principles will suffice in order to reduce GHG concentrations to the pre-industrial level. At least, humanity could then claim to have done all that is in its power to avoid triggering dangerous thresholds in natural systems.

The reflections in this chapter do not differ in practical implications from previous arguments based on the PP. However, they provide a different justification for why we should reduce our GHG emissions as strongly as possible: it is not the case that we should ambitiously mitigate in order to guarantee the best outcomes for humanity in the worst case; rather, we ought to mitigate in order to prevent an uncontrollable natural environment with potentially devastating impacts.

Acknowledgments

Earlier versions of this chapter were presented at the workshop "Ethical Underpinnings of Climate Economics," held at Helsinki University, November 2014, and at the conference "Climate Justice: Economics and Philosophy" at Cornell University, May 2016. I also enormously benefited from intensive discussions with Sebastian Cacean and Tobias Lessmeister about this chapter and the adequacy of precautionary principles in general. Moreover, I am deeply indebted to Henry Shue for his substantial comments on previous drafts and his linguistic support.

References

Barker, T. and S. Ş. Scrieciu. 2010. "Modelling Low Stabilization with E3MG: Towards a 'New Economics' Approach to Simulating Energy-Environment-Economy System Dynamics." *Energy Journal* 31: 137–64.

Betz, G. 2007. "Probabilities in Climate Policy Advice: A Critical Comment." *Climatic Change* 85 (1–2): 1–9.

Betz, G. 2010. "What's the Worst Case? The Methodology of Possibilistic Prediction." *Analyse & Kritik* 1 (1): 87–106.

Betz, G. 2015. "Are Climate Models Credible Worlds? Prospects and Limitations of Possibilistic Climate Prediction." *European Journal for Philosophy of Science* 5 (2): 191–215.

Brown, P. T., and K. Caldeira. 2017. "Greater Future Global Warming Inferred from Earth's Recent Energy Budget." *Nature* 552 (7683): 45–50.

Butler, J. H., and S. A. Montzka. 2017. "The NOAA Annual Greenhouse Gas Index (AGGI)." <https://www.esrl.noaa.gov/gmd/ccgg/aggi.html> accessed 31 January 2018.

Cox, P. M., C. Huntingford, and M. S. Williamson. 2018. "Emergent Constraint on Equilibrium Climate Sensitivity from Global Temperature Variability." *Nature* 553 (7688): 319–22.

D'Alisa, G., F. d. Maria, and G. Kallis. 2014. *Degrowth*. New York: Routledge.

Edenhofer, O., R. Pichs-Madruga, Y. Sokona, E. Farahani, S. Kadner, K. Seyboth, A. Adler, I. Baum, S. Brunner, P. Eickemeier, B. Kriemann, J. Savolainen, S. Schlömer, C. von Stechow, T. Zwickel, and J. Minx. 2014. *Climate Change 2014: Mitigation of Climate Change. Contribution of Working Group III to the Fifth Assessment Report of the Intergovernmental Panel on Climate Change*. Cambridge: Cambridge University Press.

Ellsberg, D. 2001/1961. *Risk, Ambiguity and Decision*. New York and London: Garland Publishing, Inc.

Figes, O. 1997. *A People's Tragedy: The Russian Revolution 1891–924*. London: The Bodley Head.

Friedrich, T., A. Timmermann, M. Tigchelaar, O. E. Timm, and A. Ganopolski. 2016. "Nonlinear Climate Sensitivity and its Implications for Future Greenhouse Warming." *Science Advances* 2 (11): e1501923.

Frigg, R., L. A. Smith, and D. A. Stainforth. 2013. "The Myopia of Imperfect Climate Models: The Case of UKCP09." *Philosophy of Science* 80 (5): 886–97.

Fuss, S., J. G. Canadell, G. P. Peters, M. Tavoni, R. M. Andrew, P. Ciais, R. B. Jackson, C. D. Jones, F. Kraxner, N. Nakicenovic, C. L. Quéré, M. R. Raupach, A. Sharifi, P. Smith, and Y. Yamagata. 2014. "Betting on Negative Emissions." *Nature Climate Change* 4 (10): 850–3.

Gardiner, S. M. 2006. "A Core Precautionary Principle." *The Journal of Political Philosophy* 14 (1): 33–60.

Hartzell-Nichols, L. 2012. "Precaution and Solar Radiation Management." *Ethics, Policy & Environment* 15 (2): 158–71.

Heal, G., and A. Millner. 2014. "Reflections: Uncertainty and Decision Making in Climate Change Economics." *Review of Environmental Economics and Policy* 8 (1): 120–37.

Jackson, T. 2009. "Prosperity Without Growth." Technical report, Sustainable Development Commission.

Jeffrey, R. C. 1956. "Valuation and Acceptance of Scientific Hypotheses." *Philosophy of Science* 23 (3): 237–47.

Knight, F. H. 1921. *Risk, Uncertainty and Profit*. Hart, Schaffner & Marx prize essays on economics no. 31. Boston: Houghton Mifflin.
Kriegler, E., J. P. Weyant, G. J. Blanford, V. Krey, L. Clarke, J. Edmonds, A. Fawcett, G. Luderer, K. Riahi, R. Richels, S. K. Rose, M. Tavoni, and D. P. van Vuuren. 2014. "The Role of Technology for Achieving Climate Policy Objectives: Overview of the EMF 27 Study on Global Technology and Climate Policy Strategies." *Climatic Change* 123 (3–4): 353–67.
Lenton, T. M., H. Held, E. Kriegler, J. W. Hall, W. Lucht, S. Rahmstorf, and H. J. Schellnhuber. 2008. "Tipping Elements in the Earth's Climate System." *Proceedings of the National Academy of Sciences* 105 (6): 1786–93.
Manson, N. A. 2002. "Formulating the Precautionary Principle." *Environmental Ethics* 24 (3): 263–74.
McKinnon, C. 2009. "Runaway Climate Change: A Justice-Based Case for Precautions." *Journal of Social Philosophy* 40 (2): 187–203.
McNeall, D., P. R. Halloran, P. Good, and R. A. Betts. 2011. "Analyzing Abrupt and Nonlinear Climate Changes and their Impacts." *WIREs Climate Change* 2 (5): 663–86.
Parker, G. 2013. *Global Crisis: War, Climate Change and Catastrophe in the Seventeenth Century*. New Haven, CT: Yale University Press.
Peterson, M. 2006. "The Precautionary Principle is Incoherent." *Risk Analysis* 26 (3): 595–601.
Pissarskoi, E. 2014. *Gesellschaftliche Wohlfahrt und Klimawandel: Umgang mit normativen Annahmen und Ungewissheiten bei der klimaökonomischen Politikberatung*. Munich: oekom verlag.
Rawls, J. 1971. *A Theory of Justice*. Oxford: Oxford University Press.
Sandin, P. 1999. "Dimensions of the Precautionary Principle." *Human and Ecological Risk Assessment: An International Journal* 5 (5): 889–907.
Sandin, P. 2004. "The Precautionary Principle and the Concept of Precaution." *Environmental Values* 13 (4): pp. 461–75.
Savage, L. J. 1954. *The Foundations of Statistics*. New York: Wiley.
Shue, H. 2010. "Deadly Delays, Saving Opportunities." In *Climate Ethics: Essential Readings*, edited by S. M. Gardiner, S. Caney, D. Jamieson, and H. Shue, 146–62. Oxford: Oxford University Press.
Shue, H. 2015. "Uncertainty as the Reason for Action: Last Opportunity and Future Climate Disaster." *Global Justice: Theory, Practice, Rhetoric* 8 (2): 86–103.
Smith, P., S. J. Davis, F. Creutzig, S. Fuss, J. Minx, B. Gabrielle, E. Kato, R. B. Jackson, A. Cowie, E. Kriegler, D. P. van Vuuren, J. Rogelj, P. Ciais, J. Milne, J. G. Canadell, D. McCollum, G. Peters, R. Andrew, V. Krey, G. Shrestha, P. Friedlingstein, T. Gasser, G. Arnulf, W. K. Heidug, M. Jonas, C. D. Jones, F. Kraxner, E. Littleton, J. Lowe, J. R. Moreira, N. Nakicenovic, M. Obersteiner, A. Patwardhan, M. Rogner, E. Rubin, A. Sharifi, A. Torvanger, Y. Yamagata, J. Edmonds, and C. Yongsung. 2016. "Biophysical and Economic Limits to Negative CO_2 Emissions." *Nature Climate Change* 6: 42–50.
Stainforth, D., M. Allen, E. Tredger, and L. Smith. 2007. "Confidence, Uncertainty and Decision-support Relevance in Climate Predictions." *Philosophical Transactions of the Royal Society A: Mathematical, Physical and Engineering Sciences* 365 (1857): 2145–61.

Stainforth, D. A., T. Aina, C. Christensen, M. Collins, N. Faull, D. J. Frame, J. A. Kettleborough, S. Knight, A. Martin, J. M. Murphy, C. Piani, D. Sexton, L. A. Smith, R. A. Spicer, A. J. Thorpe, and M. R. Allen. 2005. "Uncertainty in Predictions of the Climate Response to Rising Levels of Greenhouse Gases." *Nature* 433 (7024): 403–6.

Steel, D. 2013. "The Precautionary Principle and the Dilemma Objection." *Ethics, Policy & Environment* 16 (3): 321–40.

Steel, D. 2015. *Philosophy and the Precautionary Principle*. Cambridge: Cambridge University Press.

Stern, N. 2007. *The Economics of Climate Change: The Stern Review*. Cambridge: Cambridge University Press.

Stocker, T., D. Qin, G.-K. Plattner, M. M. Tignor, S. K. Allen, J. Boschung, A. Nauels, Y. Xia, V. Bex, and P. M. Midgley. 2013. *Climate Change 2013: The Physical Science Basis. Working Group I Contribution to the Fifth Assessment Report of the Intergovernmental Panel on Climate Change*. Cambridge: Cambridge University Press.

Wingspread Conference Participants (1998). "Wingspread Statement on the Precautionary Principle." <http://www.gdrc.org/u-gov/precaution-3.html>.

12

The Social Cost of Carbon from Theory to Trump

J. Paul Kelleher

12.1. Introduction

The *social cost of carbon* (SCC) is a central concept in climate change economics. My aim in this chapter is to explain the concept and then to look under its philosophical hood, so to speak. It is widely acknowledged that calculating the SCC involves making choices about the infamous topic of discount rates, and I will have much to say about discounting. But to understand the nature and role of discount rates, it is crucial to understand how each of these economic concepts—and indeed the SCC concept itself—is yoked to the concept of a *value function*, whose job is to take ways the world could be across indefinite timespans and to rank them from better to worse. A great deal, therefore, turns on the details of the value function and on just what is meant by "better" and "worse." I seek to explicate these and related issues, and then to situate them within the evolving landscape of federal climate policy in the United States.

12.2. Value Functions, Discount Rates, and the Road from Value to Policy Choice

Climate change economics employs two foundational concepts: the *feasible* and the *valuable* (Brennan, 2007). Feasibility concerns how things could be given technological constraints. For example, it focuses on what forms of conventional and environmental investment are possible, and how a given type and degree of investment would change the current trajectory of economically important phenomena.

The most salient phenomenon for mainstream climate change policy analysis is the consumption of commodities whose contribution to individual well-being is usually measured by individuals' willingness to pay for them.[1] Economists try hard to understand in detail how climate change will impact the consumption of various commodities, although the economic models I'll focus on are concerned with averages and aggregates. For example, the most simple models treat the issue of feasibility as that of identifying the range of *time paths of per capita global consumption* that are possible given various levels of financial and environmental investment. That is, if we assume that time, t, is measured in discrete periods ($t=0, 1, 2, \ldots$), and that global per capita consumption at t is an aggregate index C_t, then climate economics is concerned with which time paths (C_0, C_1, $C_2, \ldots$) are feasible, given current and future technologies.[2]

The second fundamental concept for climate economics, *value*, is typically operationalized in the form of a *value function* (this is sometimes referred to as a *social welfare function*). The job of a value function is to place feasible time paths of consumption into a ranked ordering.[3] In this way, value functions are tools for ranking the different ways the world could be.

Without yet getting into the details of any particular value function, suppose we have one that we like—call it V—which takes each feasible consumption path and assigns to it a real number. (The entire path is the V-function's "argument," we say.) For us to "like" a value function is for us to think it does a good job of using such number assignments to place the paths into a betterness ordering, with the best sequence being assigned a number higher than any other. Setting aside some key technical details concerning measurement,[4] we can use the concept of a value function to define the SCC as follows.

First, determine which feasible consumption path is the business-as-usual (BAU) path, i.e. the path that would prevail in the absence of additional climate policy. Now find the number, V(BAU), that V assigns to this BAU path. Next, determine which consumption path would result if the BAU path were changed in only the following respect: there is one extra ton of carbon

[1] This is the standard *economic* approach to measurement. Philosophers routinely criticize this way of doing things. See, e.g., Hausman (2012).

[2] In this simple framework, environmental quality is relevant only insofar as it affects aggregate consumption. It is, however, consistent with the economic foundations of climate policy analysis to treat environmental quality as having fundamental importance. For more on this, see Sterner and Persson (2008) and Heal (2009). It is also possible to switch from a focus on global per capita consumption to the per capita consumption in each of several geographic regions. See, e.g., Nordhaus and Yang (1996).

[3] Strictly speaking, a value function *represents* an ordering, which suggests that the ordering is somehow prior to the function. But I will speak as if the function *determines* the ordering, since a great deal of climate economics concerns first choosing the form and content of the value function and only then running models to discover which ordering over feasible streams results.

[4] See Broome (1991, pp. 70–5).

dioxide (CO_2) emitted in the first period. Call this the *emission-perturbed path*, or PER_E. Now find the number, $V(PER_E)$, that V assigns to PER_E. Next, subtract $V(PER_E)$ from V(BAU) to find the V-based measure of the difference made by one extra ton of CO_2 emissions (where the BAU consumption path is the comparison baseline). Now use a parallel calculation to find the difference made by adjusting the BAU path in only the following respect: reduce the first time period's per capita consumption by one dollar. Call the result the *consumption-perturbed path*, or PER_C, and calculate the difference it makes (again using BAU as the baseline) by finding V(BAU) – $V(PER_C)$. The formula for calculating the SCC is then

$$SCC = \frac{V_{BAU} - V_{PER_E}}{V_{BAU} - V_{PER_C}} \tag{1}$$

In words, the SCC is the ratio of the V-based difference made by a marginal unit of present emissions to the difference made by a marginal unit of present consumption. This ratio expresses the marginal impact on V that present CO_2 emissions make in terms of the marginal impact that present consumption makes. It is a way of making these impacts comparable, so that (for example) the benefits of greenhouse gas abatement can be compared with the monetary costs associated with abatement.

I have focused here on the SCC for the first time period, which expresses the difference made (across time) by marginal emissions in the first period in terms of the difference made by marginal consumption in that period. But the SCC can in principle be calculated for any designated time period. We will soon see why the SCC in all time periods is of special importance to climate economics.

When it comes to the SCC, clearly a great deal turns on the form of the value function, V, and on its parameter values. As regards form, the standard framework in climate economics stems from an approach set out by Frank Ramsey in 1928 (Ramsey, 1928). Ramsey's approach was utilitarian in two senses. First, Ramsey's value function had the following utilitarian form:

$$V_{Ramsey} = \sum_{t=0}^{MAX} N_t \cdot u(c_t) \tag{2}$$

where $u(c_t)$ expresses the utility of per capita consumption level c at t and N_t is the number of people alive at t. Ramsey's approach was therefore to rank consumption paths in terms of the sum total of utility they contain.[5] The second sense in which Ramsey's framework was utilitarian is that he held that the *betterness ordering* given by a utilitarian V is *ipso facto* a

[5] This not fully accurate. For technical reasons stemming from problems with the idea of an infinite sum, Ramsey sought to minimize the difference between actual realized utility and a maximum possible amount he defined in terms of a concept he called *bliss*.

rightness ordering, where a rightness ordering ranks feasible paths in terms of *everything* that matters to policy choice. Let us refer to such a tight relationship between value (as defined by a given *V*-function) and choice using the following phrase:

$$\textit{Value} \rightarrow \textit{Choice} \tag{3}$$

Many philosophers will be comfortable with (and all philosophers will be familiar with) the idea—which is just the rejection of (3)—that value considerations cannot *by themselves* fully determine proper choice from among feasible policy options. Climate economists, in my experience, find this a harder idea to make sense of.[6] I'll return to it shortly.

As already defined, the SCC is a *marginal* concept: it is the ratio of the difference made by a *very small* increase in CO_2 emissions to the difference made by a *very small* decrease in first-period consumption. Because of these marginality assumptions, equation (1) is not the only formula one can use to calculate the SCC, and Ramsey once again helps us see why.

Ramsey in effect showed that when the value function takes the following *generalized utilitarian* form:

$$V_{Ramsey} = \sum_{t=0}^{MAX} N_t \cdot u(c_t) \cdot \frac{1}{(1+\delta^t)} \tag{4}$$

where δ represents the *utility discount rate*, one can derive from *V* the following formula:

$$\rho_t = \delta + \eta \cdot g_t \tag{5}$$

Here the utility discount rate, δ, is the rate at which a given increment of utility declines in value (as measured by its impact on *V*) as its enjoyment is delayed by one period. (Ramsey believed this parameter should be set to zero to yield equation (2), which is a special case of equation (4).) η signifies the *elasticity of the marginal utility of consumption*, which represents the "curvature" of the utility function $u(\cdot)$, or the degree to which enjoying more and more consumption yields less and less utility. Finally, g_t is the rate of growth in per capita consumption between two periods, t and $t+1$, along a given time path of consumption. What Ramsey showed is that for any given time period along any given path of consumption, one can use equation (5) to derive that period's *consumption discount rate* (ρ_t). A consumption discount rate is a *rate of indifference*: for example, period 0's consumption discount rate it is the rate of return required in period 1 in order for *V* to be indifferent between investing a marginal unit of consumption in period 0 and someone's consuming it in

[6] This separates climate economists from health economists, since the latter are familiar with the idea that economists' focus—e.g. maximizing quality-adjusted life years (QALYs)—is not the only consideration relevant to health policy.

period 0. Thus if period 0's consumption discount rate is 0.05, then a 5 percent return is required in period 1 in order for V to be indifferent between consuming a marginal unit in period 0 and investing it instead.

Consumption discount rates offer a sort of shortcut around using equation (1) to calculate SCC values. So long as the incremental emissions associated with the SCC concept create incremental consumption changes in future time periods that are genuinely marginal[7] (when compared with the BAU consumption path), we can use equation (5) to derive each period t's consumption discount rate. We then define period $t+1$'s *consumption discount factor* as $\frac{1}{(1+\rho_t)}$. By multiplying each period's incremental change in consumption by its consumption discount factor, we can express period $t+1$'s consumption in terms of its period t equivalent. Then, once we have a full time path of consumption discount factors, we are in a position to translate incremental consumption in any period into its equivalent in another period. This enables us to use a time path of consumption discount factors to express a time stream of incremental consumption changes as a "present value," that is, as the change in period 0 consumption that is equivalent to it. In this way, we can avoid having to use equation (1) to calculate an SCC value: so long as the relevant consumption changes are truly marginal, we can use the consumption discount factor method instead.

Equation (5) is sometimes called the *Ramsey rule*,[8] but that name is also frequently given to a conceptually quite distinct proposition. To see this, note that equation (5) can be derived entirely from equation (4). That is, (5) gives us the marginal rate of indifference *with respect to the value considerations embodied in V*. Nothing in the manner of claim (3)—i.e. the claim that V reflects all considerations relevant to policy choice—is needed to derive (5). However, if one embraces *both* (3) *and* (4), then one can derive an ostensibly similar but entirely new proposition that is also frequently dubbed the Ramsey rule:[9]

$$r_t = \delta + \eta \cdot g_t \tag{6}$$

Here r_t represents the social rate of return on investment, or the *productivity of capital*. If one subscribes to (3) and (4), then (6) follows for reasons that, once again, Ramsey was the first to articulate. The idea is intuitive. If at a certain time, and along a certain BAU consumption path, one's consumption discount rate (as given by (5) and the right-hand side of (6)) is below the productivity of capital, then one can increase value (as measured by V) by

[7] See Dietz and Hepburn (2013) for the requirements of genuine marginality.

[8] See for example Johansson and Kriström (2015, p. 67), Adler and Treich (2015, p. 18), and Kolstad et al. (2014, p. 229).

[9] See for example Dasgupta and Heal (1979, pp. 296–7), Stern (2014, p. 456), and Arrow et al. (1996, p. 134).

investing a bit more today in order to earn a rate of return that surpasses one's *V*-based rate of indifference. Since we are supposing for the sake of argument that one subscribes to claim (3), one will apply Ramsey's logic right up to the point where one's consumption rate of indifference is exactly equal to the productivity of capital. The same logic will apply, *mutatis mutandis*, if the productivity of capital is currently lower than the consumption rate of indifference. In that case, one maximizes *V* by increasing consumption today and investing less, right up until the point at which the world conforms to the equality expressed in (6).

As noted, (5) and (6) are each frequently called "the" Ramsey Rule. It would be useful to have different names for them, and if it were up to me I'd call (5) the Ramsey *formula* and (6) the Ramsey *rule*. (Even then, I wouldn't want readers to assume that the Ramsey rule is a morally or even economically *sound* rule—only that it is a rule that many economists, including Ramsey, endorse.) In any case, the important point for now is that these express distinct propositions and that one will embrace the rule embodied in (6) only if one also embraces the tight connection between value and choice expressed in (3).

To illustrate these conceptual points, consider the view taken by environmental economists David Pierce, Edward Barbier, and Anil Markandya (PBM).[10] PBM claim that the consumption discount rate given by (5) should reflect "current generational orientated considerations."[11] By this they mean, for example, that the utility discount rate should reflect the rate that individuals today reveal in their behavior when they trade-off utility within their own lives. Since people seem to show an intra-personal bias in favor of earlier utility over later utility, PBM's approach to (5) has the implication that climate policy analysis should discount the interests of future generations in favor of those of the current generation. That is actually the standard view in climate economics, and I'll discuss it further later in the chapter. Yet to this standard view PBM add an unconventional proviso: they claim that one should first "define the rights of future generations"—decide what levels of consumption they are morally *entitled* to, say—and only then maximize a "current generation oriented" version of *V subject to the constraint that future generations' rights are fully respected*.[12] PBM are, therefore, likely to reject (3), since they do not believe that *V* captures all of the normative considerations relevant to policy choice. They would therefore also reject (6). But they needn't—and in fact shouldn't—reject (5), since (5) is entailed by any *V*-function having the generalized utilitarian form (4), which PBM's current generation oriented *V*-function does. Equation (5) merely gives the marginal rate of indifference

[10] Pearce et al., 1990. [11] Pearce et al., 1990, p. 46. [12] Pearce et al., 1990, p. 46.

between consumption in different time periods *as judged by V*. This rate of indifference is *relevant* to choice just in case the ranking given by V is relevant to choice. But that ranking can be relevant to choice without its being the *sole* relevant consideration. In contrast, the Ramsey rule expressed by equation (6) entails that the ranking given by V is the sole policy-relevant consideration.

As I've said, PBM are in the minority in rejecting an airtight relationship between value (as expressed by V) and policy choice. Most economists working in climate economics think that it is the job of economists to find a V-function that can sustain a recommendation of the form, "Identify the feasible consumption path that maximizes V and then implement the portfolio of policies required to put the world on that path."[13] But it is absolutely essential to see that this default position is entirely discretionary: absolutely nothing *in the economics* dictates that economists must embrace this particular relationship between proper policy choice and the conception of value embodied in the V-function. *V-functions merely place consumption paths into an ordering*. What that ordering signifies morally, and what public policy is to do with that ordering, are distinct and further questions.[14]

I hope it is now clear how the concept of the social cost of carbon is yoked to the idea of a value function, and how the value function might or might not be related to the practical task of policy choice. If there is indeed a gap between the policies that V ranks highest and the policies that we should pursue all-things-considered, then the SCC will not be an ironclad guide to policy choice. But if there is no gap, then the SCC will be a very useful policy instrument. For whenever the SCC exceeds the cost we would have to bear today to abate a ton of CO_2, we know that we would increase V by abating an additional ton. We also know (and this is a standard result in climate economics) that along the path that V ranks highest—what those who embrace (3) would call the "optimal" path—the SCC for each period will be equal to the cost of abating one more ton of CO_2 in that period. (This last "optimality condition" is used by economists to help work out what the optimal path looks like.) Finally, we also know that the SCC along the highest-ranked path gives us the "optimal tax," that is, the tax that would bring us on to the highest-ranked path if we applied it to each and every ton of CO_2 that is emitted around the world.

[13] Or, if they insist on restricting themselves to *climate* policy only, their recommendation will take the form, "Identify the feasible consumption path that maximizes V, then identify the associated time-path of greenhouse gas emissions (or emissions reductions), and then implement the portfolio of policies required to put the world on that path of emissions (reductions)." This is what von Below, Denning, and Jaakkola call the "naïve implementation" of the Ramsey Rule. See von Below et al. (n.d., pp. 26–7).

[14] See (Kelleher 2017a) for an extended discussion on this point.

12.3. Investigating Value Functions Further

Not all climate economists embrace a *V*-function falling into the generalized utilitarian family captured by (4). For example, in a recent book Humberto Llavador, John Roemer, and Joaquim Silvestre (LRS) explore and defend just the second half of the two-pronged view championed by Pearce, Barbier, and Markandya. In LRS's view, climate economists should reject all *V*-functions that resemble (4) and from which the Ramsey formula (5) can be derived.[15] They argue instead that the entire framework should be built around a *V*-function that ranks consumption paths with respect to how the worst-off generation fares.[16] On such a view, larger benefits for better-off generations cannot justify smaller losses for worse-off generations, *unless* the trade-off is part of a policy that somehow improves the situation of the worst-off generation. But that means LRS-style *V*-rankings are not interested in adding time-stamped gains and losses and weighing them against one another, and this in turn means that they have little use for the SCC concept as I have defined it. For it is only in the context of what economists call *additive* *V*-functions that the SCC becomes a relevant concept both from the standpoint of economics and—if policy cares about the *V*-function—from the standpoint of public policy.

V-functions (2) and (4) are not the only ones that display additivity, however. A still different one is the *prioritarian* *V*-function:

$$V_{prior} = \sum_{t=0}^{MAX} g\Big(u(c_t)\Big) \tag{7}$$

Like (2) and (4), a Ramsey formula specifying the consumption rate of indifference can be derived from (7), though it looks somewhat different from (5) since (7) sums not discounted utilities, but utilities that have been "transformed" by a function $g(\cdot)$ that gives less and less weight to extra utility the more beneficiaries already have of it.[17] But like (2) and (4)—and quite unlike LRS's "maximin" *V*-function—the additive prioritarian *V*-function and its associated Ramsey formula can be used to calculate a (prioritarian) SCC using the formula given by (1).[18]

These three families of value functions—generalized utilitarian, prioritarian, and maximin—share the feature that they address a heterogeneous intra and intertemporal demography head-on, and calibrate their parameters with an eye toward yielding betterness orderings that acknowledge trade-offs between distinct people (or at least different groups of people—e.g. generations). LRS call approaches with this feature *ethical observer* approaches, while others

[15] Llavador et al., 2015, p. 205.

[16] Actually, LRS's view is at once more complicated and more lenient than this, but I will not get into the details of their view here.

[17] Adler and Treich, 2015, p. 296. [18] Adler et al., 2017.

call them *social planner* approaches. They stand in contrast to what are called *representative agent* models, which aim to make consumption path-ranking exercises tractable by giving the societal *V*-function the same form as an anonymous currently living individual's *V*-function. Economists like this representative agent approach because there are observational data from individual behavior they can use to calibrate the value parameters, and because they are already trained to think systematically about the structure and properties of individual preference.

One significant problem for representative agent models is that they cannot adequately reflect prioritarian considerations. Prioritarian value functions like (7) rank consumption paths by applying a "concave transform," $g(\cdot)$, to each period-specific utility and then summing the results of these operations across the relevant time horizon. The transform is used to give more weight to utility improvements that accrue to those who start with less of it. The resulting conception of betterness is therefore conceptually distinct from a conception on which betterness is represented by the sum of utilities. The problem for representative agent models—models that treat societal betterness rankings as if they were the ranking of an individual—is that it is definitional of the idea of (von Neumann–Morgenstern) utility that betterness for an individual is represented by summing utilities, and not by summing concave transformations of utilities.[19] Many representative agent proponents try to evade this objection by allowing prioritarian considerations to influence η, the value parameter representing the elasticity of the marginal utility of consumption. As noted earlier, η reflects the curvature of the utility function and thus the degree to which adding more and more consumption brings less and less utility. By increasing η, representative agent proponents hope to express the *combined* curvature of an explicit utility function and an implicit prioritarian transform.[20] Yet not only is this solution conceptually quite imperfect, but it also ignores two different technical issues—explained in note 21—that absolutely require the explicit decomposition the representative agent proponent was hoping to avoid.[21]

[19] For more on this point, see Greaves (2015). Traeger (2010) derives a representative agent framework that obeys the von Neumann–Morgenstern axioms and that also permits concave transformations of utility. But these transformations are an expression of risk aversion and are relevant only in contexts of uncertainty; they drop out when certain consumption paths are being evaluated. By contrast, prioritarianism assigns values to certain consumption paths by summing concave transformations of period-specific utilities. So Traeger's framework still cannot accommodate prioritarian considerations.

[20] See Stern (1977, pp. 241–2), Dasgupta and Heal (1979, pp. 279–80), and Kaplow and Weisbach (2011) for rare exceptions in the literature that make this prioritarian transform explicit.

[21] First, "implicit prioritarian" approaches work with a utility function that is unique up to positive affine transformations, whereas no prioritarian value function is invariant to positive affine transformations of the utility function. What a prioritarian value function requires is a utility function that is invariant to *ratio* transformations. See Adler and Treich (2015, §3.2).

Whether they adopt a representative agent or a social planner approach to *V*-functions, most economists working in climate change economics rely heavily on *revealed preference* to calibrate their *V*-function's value parameters. Some go so far as to say that if parameters are *not* set in this "descriptivist" way—for instance, if they are instead set using "prescriptivist" *a priori* moral reasoning—then the whole path-ranking exercise would no longer fall within the discipline of economics.[22] Here, for example, is Martin Weitzman:

> [C]line and Stern are soulmates in their *cri de coeur* justifying $\delta \approx 0$ by relying mostly on a priori philosopher-king ethical judgements about the immorality of treating future generations differently from the current generation—instead of trying to back out what possibly more representative members of society than either Cline or Stern might be revealing from their behavior is their implicit rate of pure time preference. An enormously important part of the "discipline" of economics is supposed to be that economists understand the difference between their own personal preferences for apples over oranges and the preferences of others for apples over oranges. Inferring society's revealed preference value of δ is not an easy task in any event (here for purposes of long-term discounting, no less), but at least a good-faith effort at such an inference might have gone some way towards convincing the public that the economists doing the studies are not drawing conclusions primarily from imposing their own value judgements on the rest of the world. (Weitzman, 2007, p. 712)

Pace Weitzman, revealed preference approaches face several serious objections from the standpoint of policy analysis. First, they violate the Humean dictum that one cannot infer an "ought" from an "is." The data that revealed preference advocates wish to use to calibrate the value parameters reveal the trade-offs that real-life agents *do in fact* make in their daily lives. But, arguably, the *V*-function required for social policy analysis should reflect societal judgments about how intergenerational trade-offs *should* or *ought* to be made. It would be one thing if revealed preference advocates warned, as PBM did earlier, that such *V*-functions capture just *one among many other* considerations relevant to policy choice. But a great many economists who prioritize revealed preference take the results of their models to indicate what *proper* policy *should* look like. Few say what they would need to say to properly heed Hume's dictum: "The ranking of consumption paths I offer is more the product of

Second, integrating prioritarianism with decision theory requires an explicit choice between what Adler and Treich call *ex ante* and *ex post* versions of prioritarianism under uncertainty. Although each of these has serious theoretical costs—*ex ante* prioritarianism violates the "sure thing principle," *ex post* prioritarianism violates the *ex ante* Pareto relationship between individual betterness rankings and social betterness rankings—any bona fide prioritation must choose between them. (See Adler, 2012, ch. 7.) But implicit prioritarianism standardly leads economists who embrace it to evade the issue altogether.

[22] See Kelleher (2017b) and Kelleher and Wagner (n.d.) for more on the so-called descriptivist/prescriptivist divide in climate economics.

quantitative sociology or anthropology than it is of political theory or applied ethics. So please do not assume that the ranking I end up with is indicative of *morally justifiable* social choice." Instead, revealed preference proponents describe the results of their analyses as determining "optimal policy," and they criticize policies that would deviate from the "optimum" so-defined.[23]

Even if one grants that "current generation oriented" observations are *morally relevant* to policy analysis, a second problem with the revealed preference approach is that it is arguably predicated upon the wrong observational data. For the value parameters in revealed preference models are derived from the value parameters in individuals' own, largely self-interested consumer behavior. But the path-ranking exercise relevant to climate policy analysis concerns ranking paths (and thus policies) that will impact billions of people for centuries to come. Referring to the types of consumer behavior that revealed preference advocates wish to use to set the value parameters, Nicholas Stern writes: "as this borrowing and lending take place through private decisions... this does not necessarily answer the relevant question... namely, how do we, acting together, evaluate our responsibilities to future generations."[24] This is a strong argument against policy analysis that relies heavily upon the data standardly used to calibrate value parameters in revealed preference models. Different sorts of data could be collected, for example in the context of focus groups, but to my knowledge this work has not been done on a scale sufficient for policy analysis.[25]

The reliance (or not) on revealed preference will inevitably influence how a given value function trades off well-being at one time with well-being at another. In addition, note that whether one views a given *V*-function's trade-offs as defensible will also have much to do with one's prior stance on claim (3). That is, if one begins with the view that a value function's ranking of feasible consumption paths should be an *all-things-considered* rightness ranking, then one will seek a value function that incorporates all normative considerations one deems relevant to policy choice. However, if one instead uses a value function to rank paths in terms of *some but not all* normatively relevant considerations, then the trade-offs embodied in the value function need not coincide perfectly with the full set of trade-offs one ultimately wants embodied in policy. Consider the following distinction between moral principles described by John Broome:

> Very roughly, our moral duties can be divided into duties of beneficence—promoting good—and duties of justice... The need for this division can be illustrated by an example that comes from Judith Jarvis Thomson. A surgeon has in her

[23] See, for example, Nordhaus (1994, pp. 79ff.). [24] Stern, 2010, p. 51.
[25] For a proof-of-concept example of what this could look like, see Frederick (2003).

> hospital five patients, each in need of a different organ for transplant. One needs a new heart, one needs a new liver, and so on. Each will die unless she gets a new organ. The surgeon kills an innocent visitor to the hospital and distributes her organs to her patients. Thereby the surgeon saves five lives at the expense of one. She successfully promotes good. Yet she clearly acts wrongly... It must be that there is some principle of morality that can conflict with the duty to promote good, and is sometimes important enough to override it. In this particular case, it is evidently a duty not to inflict harm on someone, even for the sake of greater general goodness. I take this to be a duty of justice.[26]

Elsewhere Broome claims that the value function relevant to the economics of climate change is insensitive to justice-based considerations and instead aims to rank consumption paths solely in terms of their capacity to promote general goodness.[27] But if one did not understand that this is Broome's focus, one might mistakenly criticize his claims about the value parameters; after all, the parameters relevant to a path-ranking exercise concerned solely with what Broome calls "goodness" may well diverge from the parameters relevant to an exercise concerned with both goodness and justice. If one did not realize that Broome is focused on goodness only, and if one instead assumed he was focused on *all* relevant normative considerations, then one might be inclined to pick a misguided theoretical fight with him about his preferred value parameters. This in fact is my diagnosis of a great number of "disagreements" between philosophers and climate economists.[28]

The key takeaway for our purposes is that since the SCC depends so heavily on the *V*-function, it could be a giant mistake—both analytically and dialectically—to speak of "the" SCC. For if it makes sense in different contexts to be interested in different kinds of path-ranking exercise, and if each of these employs an additive *V*-function, then there will be a distinct SCC concept tied to each sort of exercise. This is especially important in the context of policy, since only an SCC tied to a *V*-function capturing all relevant normative considerations should be used to identify policies that are "optimal" in the sense relevant to final social choice.

The observations made in this section can be used to analyze the approach taken by the controversial *Stern Review* of the economics of climate change.[29] Stern adopted a value function in line with (2), with a utility discount rate of

[26] Broome, 2016, p. 921. The Fifth Assessment Report of the Intergovernmental Panel on Climate Change (IPCC) was the first IPCC report to reference and incorporate this distinction between goodness and justice. Broome was a lead author of the relevant section. See Kolstad et al. (2014, p. 215).

[27] Broome, 2012, ch. 6.

[28] I develop this analysis of prominent debates in the literature in Kelleher (2017a) and Kelleher (2017b).

[29] Stern, 2007.

essentially zero.[30] Stern's value function was not of the representative agent variety because he conceived of the utility discount rate as an explicitly moral parameter serving to weight utility gains going to distinct generations. Yet in line with more mainstream analyses—including just about all representative agent analyses—Stern chose to calibrate the elasticity of the marginal utility of consumption on the basis of revealed preference, by drawing on studies of revealed consumer behavior.[31] At the same time, Stern notes that his *V*-function "has no room . . . for ethical dimensions concerning the processes by which outcomes are reached," and that it thereby ignores "different notions of ethics, including those based on concepts of rights, justice and freedoms."[32] Stern's *V*-function was therefore in line with the goodness-focused *V*-function described and employed by Broome, and indeed Broome was acknowledged in the *Stern Review* for being an influential advisor.[33] And yet, despite working with a *V*-function that self-consciously ignores at least one set of morally crucial considerations—considerations that Stern himself seems to think are relevant to policy—Stern's headline conclusion was that "prompt and strong action is clearly warranted."[34] This is what we would expect from someone who subscribes to (3).

One can therefore find in the *Stern Review* traces of many of the approaches to the *V*-function canvassed in this section, but without a methodological discussion of whether these elements are compatible with one another, or why definitive policy conclusions can be derived from *V*-functions (and thus SCCs) that reflect only a subset of policy-relevant considerations. In flagging this I do not mean to reject the *Stern Review*'s ultimate policy conclusions. I mean only to suggest that future economic analyses of climate change will be all the stronger if they attend to the issues, distinctions, and tensions raised in this section.

12.4. The Social Cost of Carbon: From Theory to Trump

With the express aim of informing climate policy in the United Kingdom, the *Stern Review* was commissioned by the Chancellor of the Exchequer in 2005, and was released in 2006. By contrast, it was not until 2007 that the US Supreme Court compelled the Environmental Protection Agency (EPA) to

[30] Stern actually adopted a value function with a very low positive utility discount rate, but that rate was used to reflect an exogenous risk of extinction. His approach was consistent with a utility discount rate of zero, so long as extinction risks can be captured in some other way (as indeed they can be).

[31] See Stern (2007, p. 52, n. 10). Elsewhere in the *Review* Stern suggests that this is actually a flawed approach to the elasticity of the marginal utility of consumption. See, for example, pp. 33–4, box 2.1, and p. 34, n. 6.

[32] Stern, 2007, p. 32. [33] Stern, 2007, p. 32. [34] Stern, 2007, p. xv.

regulate greenhouse gas emissions as part of its responsibilities under the Clean Air Act.[35] One year later, the Ninth Circuit Court of Appeals declared that federal agencies could no longer fail to put a price on carbon emissions, holding that while there may be a range of reasonable numbers for the SCC, "the value of carbon emissions reductions is certainly not zero."[36] Partly as a result of these rulings, the Obama administration in 2009 formed the Interagency Working Group on the Social Cost of Carbon (IWG), which issued its first set of SCC figures in 2010.[37] These figures were later updated in 2013 and 2015, and were used by the Obama administration's EPA to underwrite its Clean Power Plan.[38] However, on March 28, 2017, President Donald Trump issued an executive order that disbanded the IWG, directed federal agencies not to use IWG's SCC results, and instead directed agencies to rely on 2003 guidance (OMB Circular A-4) issued by George W. Bush's Office of Management and Budget.[39] Most recently, an expert panel of the National Academies of Sciences, Engineering, and Medicine (NAS), responding to a previous request for guidance by the IWG, issued a major report on how the US's SCC values could be updated and improved in light of developments in the peer-reviewed literature.[40]

The IWG approach improved upon the *status quo ante* in 2010 by projecting SCC values for the year 2015 between $4.70 and $64.90, with a "central value" of $23.80 (in 2007 dollars). In its 2015 update, the corresponding range became $11 to $105, with a central value of $36. Much of the IWG's methodology was in line with what I have described in this chapter. Employing three of the leading "integrated assessment models" (IAMs), the analysis began with descriptions of the "business-as-usual" (BAU) consumption path and then "pulsed" that path with one ton of CO_2. Next, the models estimated the temperature increases consequent on the pulse, and then the annual decreases in consumption (i.e. "climate damages") consequent on the increase in temperature. The resulting stream of incremental damages was then discounted back to the year in which the pulse occurred in order to arrive at that year's SCC value.

In the IWG's analysis, each IAM began with five possible BAU paths. Each path was then pulsed with an extra ton of CO_2 10,000 times, with each of these 10,000 "model runs" selecting at random from a probability density function expressing climate science's best understanding of the complex relationship between greenhouse gas concentration and temperature increase. This exercise yielded 150,000 streams of climate damages (3 IAMs × 5 BAU paths × 10,000 runs), each of which was then discounted back to present

[35] 127 S Ct 1438 (April 2, 2007).
[36] *Center for Biological Diversity vs. NHTSA*, 538F.3d 1172, 1200; 9th Cir., 2008.
[37] Interagency Working Group on Social Cost of Carbon, 2010.
[38] US Environmental Protection Agency, 2015. [39] White House, 2017.
[40] National Academies of Sciences, Engineering, and Medicine, 2017.

values using the three constant discount rates of 2.5 percent, 3 percent, and 5 percent. The 150,000 SCC values for each discount rate were then averaged to yield the IWG's SCC value for that discount rate, and a fourth SCC value (the upper end of the ranges noted earlier) was produced by reporting the SCC value at the 95th percentile using the central 3 percent discount rate.[41]

The follow-up NAS report rightly criticized the IWG for using constant discount rates instead of constructing discount rates using the Ramsey methodology discussed in Section 12.2.[42] Because it is derived from the conceptually fundamental *V*-function, the Ramsey approach is theoretically appropriate and allows for the possibility (unlikely in the short term, but possible in the very long term) that consumption discount rates become negative due to the severe damages imposed by unmitigated climate change. (Consumption discount rates will be negative if g_t in equation (5) is negative and if the absolute value of the product of g_t and η is greater than the utility discount rate.) Negative discount rates have the effect of placing *more* weight on future outcomes (relative to equivalent outcomes in the present).

Both the IWG and NAS adopt the revealed preference approach to value functions discussed in Section 12.3. In defending its highest constant consumption discount rate (5 percent), the IWG argues that "many individuals smooth consumption by borrowing with credit cards that have relatively high rates... [T]he high interest rates that credit-constrained individuals accept suggest that some account should be given to the discount rates revealed by their behavior."[43] And although the NAS rejects the IWG's use of constant discount rates, the NAS agrees that (certainty-equivalent) consumption discount rates should be "consistent, over the next several decades, with consumption rates of interest"—that is, they should be consistent with the consumption rates of indifference that individuals reveal in their everyday consumer behavior.[44] The IWG and NAS therefore both face the objections to revealed preference approaches set out in Section 12.3.

Nevertheless, the IWG and NAS approaches might very well be superior to the approach we are likely to see from the Trump administration. While OMB Circular A-4 was not written with climate economics in mind, its guidance suggests that streams of incremental climate damages should be discounted at two different rates: 3 percent and 7 percent. The 3 percent value, like the IWG's central value of 3 percent, results from a revealed preference approach that calibrates society's consumption discount rate by looking to the rate at

[41] Interagency Working Group on Social Cost of Carbon, 2010.
[42] National Academies of Sciences, Engineering, and Medicine, 2017, pp. 18, 169. See Kelleher and Wagner (forthcoming) for a further important correction to the NAS report related to the Ramsey discounting methodology.
[43] Interagency Working Group on Social Cost of Carbon, 2010, p. 19.
[44] National Academies of Sciences, Engineering, and Medicine, 2017, p. 180.

which actual individuals are willing to trade present consumption for future consumption (what the NAS calls the "consumption rate of interest"). As noted, the Ramsey-inspired approach is theoretically preferable, but at least in its focus on a 3 percent discount rate OMB Circular A-4 hews closely to the Obama administration's IWG's central value. Things change, however, with Circular A-4's introduction of a 7 percent discount rate.

The 7 percent rate is meant to represent the *opportunity cost of capital*, or the average rate of return earned by private investments. Let us set aside the question of whether the correct rate for this is 7 percent and ask: why should one use the opportunity cost of capital as a discount rate? The answer will be familiar from private investment decisions. If one is choosing between investments A and B, and investment A yields a 6 percent return and B yields 7 percent, then one should choose B—other things being equal and assuming the upfront costs of each investment are the same. A different but equivalent way to arrive at this conclusion is to discount A's future payout at the rate of return offered by the best alternative investment (in this case B), and then to subtract from this discounted value the initial cost of the investment. If the resulting value is negative, that signifies that the best alternative investment (B) is better than the investment under investigation (A). In recommending 7 percent as an acceptable discount rate, OMB Circular A-4 is essentially directing federal agencies to keep an eye on the opportunity costs—i.e. the best alternative investments—of federal projects.

Even if 7 percent is a reasonable estimate of the long-term rate of return on private investment (and there are good reasons to doubt that it is),[45] it is widely agreed in the theoretical literature that this is the wrong way to take opportunity costs into account, especially for projects having long-term effects. The superior way, as indeed Circular A-4 itself notes, is to view the benefits of foregone projects as a *cost*, rather than as a *discount rate*.[46] To the extent that investment in project A displaces project B, project B's annual payouts should be deducted from the annual payouts of A. To work out B's payouts, one will of course have to take into account the rate of return one could have received if one had invested in B instead: first one determines the incremental improvements in consumption made possible by B, and then one discounts those back to a single present value using consumption discount rates, which, as we've seen, Circular A-4 sets at 3 percent. So instead of featuring in the analysis as a discount rate, B's 7 percent rate of return should be used to calculate the consumption gains one foregoes by investing in A instead. As Dasgupta, Sen, and Marglin put it in 1972, the productivity of capital is relevant because "the present value of the consumption stream

[45] Council of Economic Advisors, 2017. [46] Office of Management and Budget, 2003, p. 33.

forgone when such investment is displaced... is relevant, and this present value is relevant as a cost, not as a discount rate."[47]

The method advocated by Dasgupta, Sen, and Marglin carries even more weight in contexts where a project's impacts will be felt for many years to come. That is because OMB's 7 percent method asks us to consider only the upfront costs of making the capital investment. It does not consider the downstream costs associated with whatever productive activity makes the investment's payout possible in the first place. To see this, consider the example that compared A and B in the previous paragraph. Suppose again that B pays investors an annual 7 percent return and that when we discount A's future benefits at 7 percent and then subtract the upfront cost of A's investment, we get a negative number. According to Circular A-4's logic, that suggests B is the better investment. But now suppose that B is an investment in a paper company, and that while the paper company pays out stock dividends at 7 percent per year, the company's pollution creates consumption damages for third parties for decades to come. Surely a full accounting of A's and B's relative merits should take this into account, but Circular A-4's 7 percent method does not do this. Dasgupta, Sen, and Marglin's method does. For the latter directs us to compare each project by working out the *net* incremental consumption impact of each, and discounting each net impact stream at the consumption discount rate. This method, which is broadly in line with that recommended by the IWG and NAS, has the virtue of not being unduly swayed by ostensibly high rates of return for investors that involve significant costs down the line for uninvolved parties.

In the end the Trump administration leaned on OMB Circular A-4's 7 percent method to help arrive at quite low SCC values. Almost certainly fearing a judicial ruling requiring it to retain at least *some* SCC value, the Trump EPA combined the 7 percent discount rate with the view that SCC values for US public policy should take into account only *domestic* climate damages. This resulted in a "central" SCC value of $1 (it had been $51 in the last Obama-era analysis).[48] The move from a global to a domestic SCC obviously raises its own economic and ethical issues, which I lack the space to discuss here.[49] But all on its own, the use of a 7 percent discount rate was a significant mistake from the standpoint of both economics and ethics.

[47] Dasgupta et al., 1972, p. 171.

[48] See Armstrong (2017). These figures are denominated in 2017 dollars, whereas the final Obama-era analysis used 2007 dollars.

[49] I know of no extensive philosophical discussion of this move. For the most comprehensive *economic* analysis I know of, see Ceronsky et al., 2014, pp. 5–12.

12.5. Conclusion

Shortly after President Trump's executive order, and in response to the recent NAS report, the venerable nonprofit organization Resources for the Future announced that it will lead "a multi-year, multidisciplinary research initiative that will advance the NAS recommendations and lead to a comprehensive update of the social cost of carbon estimates, as well as enhance the capabilities of decisionmakers and analysts worldwide who use the social cost of carbon to measure the benefits of emissions reductions."[50] This is heartening news at a time when federal climate policy analysis (and policy) has been anything but heartening. My aim in this chapter has been to assist this and related efforts by illuminating some of the theoretical and philosophical dimensions that require further investigation and discussion among and between economists and philosophers, especially (but not only) in the age of Trump.

Acknowledgments

For very helpful comments on earlier versions of this chapter, I thank Kian Mintz-Woo, Matthew Rendall, Jeff Round, and the editors of this volume.

References

Adler, M. 2012. *Well-Being and Fair Distribution*. Oxford: Oxford University Press.

Adler, M., D. Anthoff, V. Bosetti, G. Garner, K. Keller, and N. Treich 2017. "Priority for the Worse-Off and the Social Cost of Carbon." *Nature Climate change* 7 (6): 443–9.

Adler, M. D., and N. Treich 2015. "Prioritarianism and Climate Change." *Environmental & Resource Economics* 62 (2): 279–308.

Armstrong, A. 2017. "Trump Administration Drops Social Cost of Carbon from $51 to $1." *S&P Global Marketplace Intelligence*. <https://www.snl.com/InteractiveX/articleabstract.aspx?id=42286458&KPLT=8>.

Arrow, K. J., W. R. Cline, K. G. Maler, M. Munasinghe, R. Squitieri, and J. E. Stiglitz. 1996. "Intertemporal Equity, Discounting, and Economic Efficiency." In *Climate Change 1995: Economic and Social Dimensions of Climate Change: Contribution of Working Group III to the Second Assessment Report of the Intergovernmental Panel on Climate Change*, edited by J. P. Bruce, H. Lee, and E. F. Haites, 127–44. Cambridge: Cambridge University Press.

[50] <http://www.rff.org/research/collection/updating-and-improving-social-cost-carbon>.

Brennan, G. 2007. "Discounting the Future, Yet Again." *Politics, Philosophy & Economics* 6 (3): 259–84.

Broome, J. 1991. *Weighing Goods: Equality, Uncertainty and Time.* Oxford: Basil Blackwell.

Broome, J. 2012. *Climate Matters: Ethics in a Warming World.* New York: W. W. Norton & Company.

Broome, J. 2016. "The Well-Being of Future Generations." In *The Oxford Handbook of Well-being and Public Policy* edited by M. D. Adler and M. Fleurbaey. Oxford: Oxford University Press.

Ceronsky, M. et al. 2014. "Comments on the U.S. Social Cost of Carbon." <https://academiccommons.columbia.edu/catalog/ac:182531>.

Council of Economic Advisors. (2017). "Discounting for Public Policy: Theory and Recent Evidence on the Merits of Updating the Discount Rate." <https://obamawhitehouse.archives.gov/sites/default/files/page/files/201701_cea_discounting_issue_brief.pdf>.

Dasgupta, P. S. and G. M. Heal. 1979. *Economic Theory and Exhaustible Resources.* Cambridge: Cambridge University Press.

Dasgupta, P., A. Sen, and S. Marglin. 1972. *Guidelines for Project Evaluation.* New York: United Nations.

Dietz, S. and C. Hepburn. 2013. "Benefit–Cost Analysis of Non-Marginal Climate and Energy Projects." *Energy Economics* 40: 61–71.

Frederick, S. 2003. "Measuring Intergenerational Time Preference: Are Future Lives Valued Less?" *Journal of Risk and Uncertainty*, 26 (1): 39–53.

Greaves, H. 2015. "Antiprioritarianism." *Utilitas* 27 (1): 1–42.

Hausman, D. 2012. *Preference, Value, Choice, and Welfare.* Cambridge: Cambridge University Press.

Heal, G. 2009. "The Economics Of Climate Change: A Post-Stern Perspective." *Climatic Change* 96 (3): 275–97.

Interagency Working Group on Social Cost of Carbon. 2010. "Social Cost of Carbon for Regulatory Impact Analysis under Executive Order 12866." <https://obamawhitehouse.archives.gov/sites/default/files/omb/inforeg/for-agencies/Social-Cost-of-Carbon-for-RIA.pdf>.

Johansson, P.-O., and B. Kriström. 2015. *Cost–Benefit Analysis for Project Appraisal.* Cambridge: Cambridge University Press.

Kaplow, L., and D. Weisbach. 2011. "Discount Rates, Social Judgments, Individuals' Risk Preferences, and Uncertainty." *Journal of Risk and Uncertainty* 42 (2): 125–43.

Kelleher, J. P. 2017a. "Pure Time Preference in Intertemporal Welfare Economics." *Economics and Philosophy* 33: 441–73.

Kelleher, J. P. 2017b. "Descriptive versus Prescriptive Discounting in Climate Change Policy Analysis." *Georgetown Journal of Law & Public Policy* 15: 957–77.

Kelleher, J. P., and G. Wagner. (Forthcoming). "Ramsey Discounting Calls for Subtracting Climate Damages from Economic Growth Rates." *Applied Economics Letters.* <https://gwagner.com/ramsey-discounting-calls-for-subtracting-climate-damages-from-economic-growth-rates/>.

Kelleher, J. P., G. and Wagner. (n.d.). "Prescriptivism, Risk Aversion, and Intertemporal Substitution in Climate Economics." <https://gwagner.com/kelleher-wagner-prescriptivism-risk-aversion-and-intertemporal-substitution-in-climate-economics/>.

Kolstad, C., K. Urama, J. Broome, A. Bruvoll, M. Cariño-Olvera, D. Fullerton, and F. Jotzo. 2014. "Social, Economic and Ethical Concepts and Methods." *Climate Change 2014: Mitigation of Climate Change. Working Group III Contribution to the Fifth Assessment Report of the Intergovernmental Panel on Climate Change,* edited by Ottmar Edenhofer et al., 173–248. Cambridge: Cambridge University Press.

Llavador, H., J. E. Roemer, and J. Silvestre. 2015. *Sustainability for a Warming Planet.* Cambridge, MA: Harvard University Press.

National Academies of Sciences, Engineering, and Medicine. 2017. "Valuing Climate Damages: Updating Estimation of the Social Cost of Carbon Dioxide." <https://doi.org/10.17226/24651>.

Nordhaus, W. D. 1994. *Managing the Global Commons: The Economics Of Climate Change.* Cambridge, MA: MIT Press.

Nordhaus, W. D., and Z. Yang 1996. "A Regional Dynamic General-Equilibrium Model of Alternative Climate-Change Strategies." *The American Economic Review* 86 (4): 741–65.

Office of Management and Budget (2003). "Circular A-4." <https://www.transportation.gov/sites/dot.gov/files/docs/OMB%20Circular%20No.%20A-4.pdf>.

Pearce, D., E. Barbier, and A. Markandya. 1990. *Sustainable Development: Economics and Environment in the Third World.* Cheltenham: Edward Elgar.

Ramsey, F. P. 1928. "A Mathematical Theory of Saving." *The Economic Journal* 38 (152): 543–59.

Stern, N. 1977. "The Marginal Valuation of Income." In *Proceedings of the Association of University Teachers of Economics, Edinburgh Meeting of April 1976,* edited by M. J. Artis and A. R. Nobay, 209–58). Oxford: Basil Blackwell.

Stern, N. 2007. *The Economics of Climate Change: The Stern Review.* Cambridge: Cambridge University Press.

Stern, N. 2010. "The Economics of Climate Change." In *Climate Ethics: Essential Readings,* edited by S. Gardiner, S. Caney, D. Jamieson, H. Shue, and R. K. Pachauri, 39–76). Oxford: Oxford University Press.

Stern, N. 2014. "Ethics, Equity and the Economics of Climate Change Paper 2: Economics and Politics." *Economics and Philosophy* 30 (3): 445–501.

Sterner, T., and U. M. Persson. 2008. "An Even Sterner Review: Introducing Relative Prices into the Discounting Debate." *Review of Environmental Economics and Policy* 2 (1): 61–76.

Traeger, C. 2010. "Intertemporal Risk Aversion: Or Wouldn't it be Nice to Tell Whether Robinson Crusoe is Risk Averse?" Unpublished ms.

US Environmental Protection Agency. 2015. "Regulatory Impact Analysis for the Clean Power Plan Final Rule." <https://www3.epa.gov/ttnecas1/docs/ria/utilities_ria_final-clean-power-plan-existing-units_2015-08.pdf>.

von Below, David, Francis Denning, and Nikko Jaakkola. (n.d.). "The Climate Pension Deal: An Intergenerational Bargain." Unpublished manuscript. <https://www.nottingham.ac.uk/climateethicseconomics/documents/papers-workshop-2/jaakkola-et-al.pdf>.

Weitzman, M. L. 2007. "The Stern Review of the Economics of Climate Change." *Journal of Economic Literature* 45 (3): 703–24.

White House (2017). "Presidential Executive Order on Promoting Energy Independence and Economic Growth." <https://www.whitehouse.gov/the-press-office/2017/03/28/presidential-executive-order-promoting-energy-independence-and-economi-1>.

13

Long-Term Climate Justice

John Nolt

Anthropogenic global climate change resulting mainly from the burning of fossil fuels during the historically brief fossil fuel era will displace, sicken, injure, and kill huge numbers of people over the coming centuries, perhaps millennia. This is an unprecedented injustice. But explaining why it is unjust taxes the explanatory power of current theories of justice. Human rights theories do well in explaining why we should respect the rights of the global poor over the next few generations (see especially Shue, 2014; Caney, 2010). But they do less well with *long-term* climate justice: that is justice for people who will be born only after—perhaps long after—we have died. Stephen Gardiner (2011, 2009, 2003) has highlighted the difficulty for contractarian and contractualist theories in particular of accounting for the evident injustice of long-term climate damage. (In a nutshell, such theories require interaction, at least of a hypothetical sort, and that is problematic among people who are widely separated in time.) I largely agree with his critique. Here I argue that long-term climate injustice also taxes the explanatory powers of other theories of justice.

This chapter critiques the two currently most prominent accounts of climate injustice: that it is unfair intergenerational distribution of burdens and benefits, and that it is a violation of future people's rights. But, first, it is important to consider what is known of the cumulative long-term harms of anthropogenic climate change.

13.1. The Cumulative Long-Term Harms of Anthropogenic Climate Change

Greenhouse gas emissions are already killing people. Recent efforts to estimate the *annual* global mortality rate attributable to anthropogenic climate change now and over the next few decades all yield figures in the hundreds of thousands.[1] None of these estimates account for deaths resulting from attempted migration or military action attributable to climate change. It has been argued, for example, that drought induced by climate change helped to spark the Syrian civil war and the resultant migrations into Europe and elsewhere (Gleick, 2014). If such effects were considered, current casualty and mortality estimates would be higher. Worse may be in store. Military establishments of many nations regard climate change as a "threat multiplier" and are preparing for climate change-induced conflict (Huebert et al., 2012).

Climate casualties will, moreover, continue to occur long after we stop burning fossil fuels. This is a function, in part, of the persistence of carbon dioxide in the atmosphere and, in part, of long-term climate feedbacks.

A comprehensive survey of work on the persistence of carbon dioxide finds that "The models agree that 20–35% of the CO_2 remains in the atmosphere after equilibration with the ocean (2–20 centuries)" (Archer et al., 2009, p. 117). This does not mean that individual molecules remain in the air that long; it is, rather, an estimate of how long the global CO_2 concentration remains elevated.

Yet unless we alter the planet's climate by geoengineering (which has its own dangers), long-term feedbacks, nearly all of them positive, will prolong elevated temperatures for millennia after carbon emissions cease and the CO_2 concentration declines. Archer et al. (2009) conclude that

> Nowhere in these model results or in the published literature is there any reason to conclude that the effects of CO_2 release will be substantially confined to just a few centuries. In contrast, generally accepted modern understanding of the global carbon cycle indicates that climate effects of CO_2 releases to the atmosphere will persist for tens, if not hundreds, of thousands of years into the future.

Zeebe (2013) is more specific: surface temperatures for a total carbon input of 2.5 trillion metric tons over 500 years (a high but not impossible estimate), assuming a middle-of-the-road climate sensitivity estimate of 3°C for doubling of CO_2, will remain elevated for 23,000–165,000 years.

Long-term casualties will result not only from the effects of elevated temperatures—floods, wildfires, drought and consequent crop failure, severe storms, the spread of tropical diseases, inundation of coastal cities, and so on—but also from violence sparked by dislocation and competition for resources. Loss of the food productivity of the oceans due to CO_2-induced acidification is also likely to produce many deaths. The Interacademy Panel on International Issues warns that "Marine food supplies are likely to be reduced

[1] Here are some recent estimates: DARA (2012) 400,000 currently; 700,000 by 2030. WHO (2014) 250,000 during the period 2030–50 (selected causes only). Springman et al. (2016) greater than 500,000 merely from reduction in food availability by 2050.

with significant implications for food production." This too is a long-term prospect: "Ocean acidification is irreversible on timescales of at least tens of thousands of years" (IAP 2009).

The scope of these harms is unprecedented. They accumulate over millennia, but their ultimate magnitude depends largely on humanity's total carbon emissions during the fossil fuel era. Even small emissions contribute causally to all subsequent temperature elevation and hence to the resulting casualties (Nolt, forthcoming b; 2015a, pp. 14–15).

How many casualties will anthropogenic climate change produce in total? The question can meaningfully be asked only relative to a specified time period. Even at today's rates, as John Broome (2012: 33) has noted, climate change will produce tens of millions of deaths by 2100. But the mortality rate is increasing. The number of *casualties* by the end of this century—counting instances of dislocation, injury, and disease, as well as death—will be much greater. Beyond 2100 prediction becomes quite uncertain, but it is not impossible that cumulative climate casualties over the next millennium will reach into the billions (Nolt 2011b, 2013). Adaptation, mitigation, and geoengineering may help to reduce that total. Even so, cumulative casualties of the order of magnitude of the world wars are likely.

13.2. Harming Distant Future People: The Non-Identity Problem

Strangely, some theorists think that even if there are very large numbers of casualties, those occurring in the distant future, at least, will not amount to harm. They infer this from the non-identity problem. First formulated by Gregory Kavka (1982) and brought to prominence by Derek Parfit in *Reasons and Persons* (1984), the problem arises from the fact that different intergenerational choices may result in populations composed of different people, or even, over long timespans, global populations that are entirely distinct. Decisions with such different-population outcomes are known as non-identity cases (Parfit, 1984, ch. 16).

Large-scale, long-term public policy decisions are non-identity cases. Different policies create different employment patterns. People take different jobs, in different places under one policy than under another, and as a result they or their children acquire different mates. Different couplings produce different children, and these differences ramify through the generations. Thus after a number of generations under one policy the global population can be composed largely of people who would not have lived if a different policy had been chosen.

Climate or energy policies can easily produce such results. Suppose that on a business-as-usual policy people in the distant future will live in a degraded world, but their lives, though harsh, will still be worth living. On a policy of steep carbon reduction, those people will not exist at all (though, presumably, others will). If we continue with business-as-usual, then, we will (so the argument goes) not have harmed our distant posterity, even though they will live in a degraded world. We might even have benefited them; for they will live, and had we chosen otherwise they would not live at all. The crux of the argument is as follows:

1 Under business-as-usual, distant future people will not be worse off than they would have been otherwise (since otherwise they never would have existed).
2 People are harmed by an action or policy only if it makes them worse off than they would have been otherwise.
3 So Distant future people will not be harmed by business-as-usual.

Some authors have found this inference convincing. After laying out the non-identity problem Steve Vanderheiden, for example, writes

> Given the apparent impossibility of our present policy choices directly harming particular future persons, it appears that we cannot have any duties with respect to them, including negative duties not to harm them and positive duties to assist them, since neither is possible, at least insofar as these obligations are to persons. (Vanderheiden, 2008, p. 123)

Later he asserts without qualification that "we cannot harm the interests of a person that does not yet exist" (p. 129). He subsequently struggles to explain why our greenhouse gas emissions are unjust to future people. Crucial to his explanation is the human capacity for foresight:

> Because of the problematic nature of establishing obligations to future and therefore nonexistent persons, foresight instead considers present and foreseeable future duties we have, and will come to have, to actually existing persons, including those that will actually exist in the future. If we can foresee that our current actions will, in the future, cause harm to or violate the rights of some future person, or if they are likely to do so, then we should refrain from engaging in those actions. (p. 137)

Vanderheiden's explanation of the injustice is not unreasonable, but the strain is evident. If the non-identity problem shows that "directly harming future persons is impossible," then how can we foresee that our actions "will, in the future, cause harm to or violate the rights of some future person" or are "likely to do so"?

Other theorists reject the non-identity argument by denying premise 2. Some, such as Lukas Meyer, fall back on a non-comparative notion of harm. Meyer holds that a person is harmed by an action if, as a result of it, her welfare does not exceed a certain threshold of decency:

> While the person being harmed according to the threshold notion does not imply that the same person is worse off than she would have otherwise been at that time, it does imply that this person is worse off than she ought to be. (Meyer 2015)

Meyer can deny premise 2 by maintaining that distant future people in the business-as-usual scenario are harmed despite the fact that they are not worse off than they would have been otherwise, since they are worse off than they ought to be.

But one need not resort to non-comparative notions of harm to deny premise 2, for it is in fact false in a mundane comparative sense. Imagine a woman crossing a street to catch a flight at an airport, who is hit and injured by a reckless driver and thereby prevented from taking her flight.[2] Certainly the driver harms her. Now suppose that the plane crashes, killing all on board, so that, as it turns out, he has also saved her life. Then she is *not* worse off as a result of his reckless driving than she otherwise would have been. So according to premise 2 he has not harmed her. But in fact he has, as her injuries reveal.

Premise 2 is false because, as it is commonly understood by its advocates, it implicitly aggregates the salient consequences of the action (reckless driving, in this case) or policy into an overall assessment of "worse off" or "not worse off." But some consequences may be harmful while others are not harmful, or even beneficial. Harm is done if some of them are harmful, even if greater benefit ultimately results. Thus, although premise 2 is false, something similar is true: people are harmed (in a comparative sense) by an action or policy only if at least one of its consequences makes them worse off than they would have been had that consequence not occurred.

The reckless driving has two salient consequences. The first is that the woman is injured; the second is that she is not killed in the plane crash. She is worse off as a result of her injuries from the automobile accident than she would have been had that consequence not occurred. Hence the driver has harmed her in a mundane and morally relevant comparative sense. Thus the fact that his recklessness has not made her worse off overall does not entail that it has not harmed her.

The same is true in the non-identity case. Again, consider a particular example: a man, say, in the distant future of the business-as-usual scenario, who suffers a debilitating heat stroke in sweltering temperatures caused by our

[2] This is a modified version of an example introduced by James Woodward (1986).

(collective) carbon emissions. Our pursuit of business-as-usual has for him two momentous consequences: his having been born and the heat stroke. He is worse off as a result of the heat stroke than he would have been had it not occurred. Hence we have harmed him in a mundane and morally relevant comparative sense. Yet, as in the case of the woman, the fact that our emissions have not made him worse off overall does not entail that they have not harmed him.

Yet—among philosophers, at least—the tendency to define the harm of a policy or action by aggregating its consequences is remarkably persistent. Consider, for example, a recent paper by Duncan Purves (2016). Purves assumes this version of premise 2: "An event *e* constitutes a counterfactual comparative harm for a subject S if and only if S would have been better off had *e* not occurred" (p. 217). He then infers that a person cannot be harmed in a comparative sense by an action if that person would not have existed without it. Again this aggregates the consequence of the person's existence with further consequent injuries into an overall assessment that the person would not be worse off without that action.[3] Purves's view thus implies that the man who was killed by heat stroke in the previous example was *not* harmed in a comparative sense by our acts of carbon emission.[4] But how could the fact that he would not have existed without our emissions have made him invulnerable to comparative harm from them?

Consider such cases from the standpoints of those affected. Imagine that we are living in 2150. Two people, Al and Beth, both friends of ours, have just been killed at the same young age by a hurricane induced or intensified by greenhouse gas emissions. Suppose, moreover, that Al would have been born even if people back in the early twenty-first century had followed policies that would have prevented that hurricane, but Beth would not. (Maybe the science of 2150 even enables us to know this.) On Purves's view, the twenty-first century emissions that harmed Al did not harm Beth, because they were necessary for her existence. Yet Beth's birth and the hurricane were distinct consequences of the emissions. Beth *did* exist, and that made her just as vulnerable to the further effects of those emissions as Al was. CO_2 emissions harmed both her and Al in the same mundane comparative sense; one consequence (the hurricane or its intensification) made both worse off than they would have been without it. Therefore, because people in the twenty-first century knew better and failed to reduce emissions or alleviate their harmful effects, they harmed both Al and Beth in the same morally culpable way.

[3] Remarkably, he goes on to use that claim to advocate an intergenerational social discount rate. I do not consider that further argument here, since the central claim upon which it is based (premise 2) is, as he and others interpret it, false.

[4] Purves concedes that there may be morally significant non-comparative forms of harm. These, however, are not at issue in the present example.

In sum, then, the deaths and injuries inflicted upon future people by greenhouse gas emissions are harms in a morally relevant comparative sense. Our emissions harm distant future people, even those whose existence depends upon them, exactly as they harm present people—by afflicting them with injury, sickness, homelessness, or death (Nolt, 2011a, 2013, 2015b).

This conclusion incriminates all of us who burn fossil fuels. Those who nevertheless find the non-identity argument persuasive might do well to re-examine their motives, for they are not disinterested parties.

13.3. Climate Justice as Fair Intergenerational Distribution

In what sense if any, then, is the harm we are doing to future people *unjust*? There are two prominent conceptions of intergenerational justice. The first is that it is fair distribution of something. The second is that it is respect for the rights of future persons. I'll consider each in turn.

Fair distribution is certainly an aspect of justice, but in long-term climate justice its role is secondary, for two reasons: (1) there is no effective way for us to ensure fair distribution of anything across many future generations, and (2) questions of distribution pale in comparison with the magnitude and duration of the harms of climate change.

It is not possible here to survey the full range of conceptions of fair distribution. Instead, I will critique one that is particularly germane to our topic (fair intergenerational distribution of the CO_2 emissions budget)[5] and then explain how that critique can be generalized to others.

The idea of an emissions budget begins from the assumption that a certain level of warming—usually 2°C, sometimes not even that much—is permissible. (That assumption is questionable, but let's grant it here.) To have a reasonable chance of staying within 2°C, the standard wisdom is that we must not emit a total of more than about a trillion tons of carbon into the atmosphere. We have already emitted roughly half that. Therefore, the remaining budget is about half a trillion tons. The question is how to fairly distribute what remains.

That question is largely academic. The Paris Agreement notwithstanding, the infrastructure commitments of developed and rapidly developing nations and of global corporations have already spoken for much, if not all, of what remains.

Too narrow a focus on the emissions budget can also foster a certain blindness to the facts on the ground. At one point this blindness seems to befall even Darrell Mollendorf, who is well aware of those facts:

[5] For more on the idea of an emissions budget, see Mollendorf (2014, pp. 120–2), and his references to the work of Henry Shue.

> Our failure to mitigate climate change would constitute using more than our share of a morally constrained intergenerational CO_2 emissions budget; that would not be a failure to help others in need. On the contrary, it would harm people in several ways. First, it would involve taking more of the CO_2 emissions budget than rightly belongs to us. Second, it would undermine the capacity for moral choice of future persons (roughly the next generation) who would be left with the choice of either emitting CO_2 in excess of the cumulative budget—and in doing so, heaping massive risk and uncertainty upon subsequent generations—or making a rapid and human-development slowing transition to renewable energy. Third, regardless of which choice the next generation makes, we would be implicated in the subsequent suffering because we framed the choice of that generation.
>
> (Mollendorf, 2014, p. 159)

Mollendorf argues that maldistribution of the carbon budget would result in three sorts of "harms": (1) taking more of the CO_2 emissions budget than rightly belongs to us, (2) undermining the capacity for moral choice of future persons, who would either have to break the carbon budget or heap massive risk and uncertainty upon subsequent generations, and (3) the suffering subsequent to *their* choice.

But what of suffering subsequent to *our* choices? It is not just that if we exhaust the carbon budget we force upon subsequent generations the unpalatable choice of changing the energy infrastructure or doing harm. Since greenhouse gas emissions are already killing people, even consuming our alleged "share" of the carbon budget is harmful to many. That consumption, of course, also produces great benefits, but these do not automatically justify the harm. Nor is the harm merely risk. There is risk to individuals, of course, but globally we must reckon not merely with acceptable risks, but with acceptable casualties.

I have focused on just distribution of the carbon budget because of its direct relevance to our topic. But for the purposes of long-term climate justice, other proposals for fair intergenerational distribution are impaired by the same blindness. If we conceive climate justice only as fair intergenerational distribution of wealth, opportunities, natural capital, or the like, we will have failed to account for the injustice of killing and otherwise harming large numbers of innocent people.

Shall we, then, understand the injustice of contemporary carbon emissions as unfair distribution of injury, sickness, homelessness, and death? That would inject a dose of realism into our theorizing, provided that temporally distant harms are not belittled by a social discount rate.[6] But few justice

[6] Social discount rates have legitimate uses. But they are unjustifiable for moral assessment of bodily harm. For further explanation, see Nolt (2015a, sec. 4.1.3) and Nolt (forthcoming b). See also Parfit (1984, appendix F).

advocates are willing to treat injury, sickness, homelessness, and death as items for distribution.

13.4. Climate Justice as Respect for the Rights of Future People

Thus we come to the second approach to climate justice, according to which it is a matter of respecting human rights. This has evident advantages over distributional approaches: human rights theories prohibit forced homelessness, harm, and killing, and they allegedly trump practical worries that climate justice may be too demanding. They are, moreover, often regarded as universal, which suggests that all people, present and future, have equal rights.

Mollendorf, for example, maintains that three fundamental human rights discussed by Simon Caney (2010) extend to future people:

> (1) the right not to be arbitrarily deprived of one's life; (2) the right not to have others cause serious threats to one's health; and (3) the right not to have others deprive one of the means of subsistence. (Mollendorf, 2014, p. 222)

These are negative rights—specifically, rights to non-harm. They are universal if any rights are, so the case for ascribing them to future people is particularly strong.

But other rights theorists demur. Steve Vanderheiden is wary not only of the non-identity problem, but also of what I call the "non-existence problem." The latter is *not* the idea that there may be no people a century or a millennium from now, which is indeed possible, though unlikely. It is, rather, the idea that people who do not yet exist do not yet have rights. Thus Vanderheiden resists claiming that future people currently have rights. Instead he maintains that acting in a way that violates the rights they will have when they do come to exist constitutes morally culpable negligence (Vanderheiden, 2008, pp. 129–37).

This concession is unnecessary. Like virtually everyone else, Vanderheiden presupposes a Newtonian conception of time according to which the universe is divided into successive moments of a single absolute time. But Einstein and Minkowski refuted that conception over a century ago. By itself, of course, Newtonian physics does not imply the non-existence of future people. To reach that conclusion, you must also assume metaphysical presentism—the idea that everything that exists does so in a single moment, the constantly advancing universal present. But presentism is better adapted to Newtonian physics than to relativity, which has no use for a universal present. On the most straightforward understanding of relativity, future people exist in the same way we do, though in regions of space-time from which no signal can reach us.

All evidence for the non-existence of the future boils down to its inability to affect us now. But relativity accounts for that evidence without denying the future's existence. Indeed, if inability to affect us now implied non-existence, then distant supernovae whose light has not yet reached us (and which therefore cannot affect us now) would not exist. The resultant quasi-solipsism (all existence lies within my past and present—or, in relativistic terms, within my past light cone) would be absurd.

There are only two ways (barring humanity's imminent extinction) to maintain that future people do not exist: either to embrace some uneasy amalgam of relativity and metaphysical presentism[7] or to reject relativity. Both are unpromising.[8] Therefore, despite its seeming obviousness, the claim that future people do not (now) exist is far from secure. Elsewhere I have argued that it is false (Nolt, forthcoming a).

Of course, if future people do exist and have rights that we ought to respect now, that may drastically complicate human rights theory. So maybe theoretical parsimony, or constraints of practicality, require us to reject the idea that they have such rights. Or maybe, despite relativity, metaphysical presentism is somehow true. Still, Vanderheiden's point stands: the rights that future people will have when they come to exist have moral force today. In no way, therefore, does the non-existence objection refute long-term human rights theory.

Still, long-term rights theory faces two other problems. The first is its lack of a viable moral psychology. The rights of future people, however defensible theoretically, tend not to move us. In part this is because their rights, unlike those of present people, are not balanced by equivalent duties to us, since we will be dead before most of them are born. But that is not all. The very idea that future people have rights is alien to our traditions and to our apparent self-interests. Moreover, the legalistic us-against-them jargon in which theorists often describe those rights (they give future "persons" "binding claims" against us that impose potentially costly "duties" upon us) is in long-term rights theory self-defeating. It suggests that we should be *forced* to comply. But how? Distant future people have no power over us. Their few contemporary advocates have little more. And, being for the most part fossil fuel consumers themselves, these contemporary advocates are embarrassingly implicated in the rights violations they condemn. It is no wonder, then, that we are little moved by talk of the rights of distant future "persons."

The absence of a plausible moral psychology is bad. But the second problem is worse: in long-term rights theory unfulfillable duties proliferate. Mollendorf

[7] I have in mind the work of sophisticated metaphysical presentists, who are well aware of the constraints of relativity. See especially Zimmerman (2011).

[8] See Maudlin (2011, esp. pp. 255–9). Also relevant are Mozersky (2011) and Savitt (2011).

(2014), who holds that future people do have rights that bind us now, puts it this way:

> Given current trends in global inequality and poverty, the costs of mitigating climate change might seem most likely to fall on the poor with the predictable effect that their rights—not to be arbitrarily deprived of life, health and the means of subsistence—will be violated. Suppose that pessimistic prediction about climate change mitigation were correct; in that case, it would be unclear how the human rights approach gives a reason to prefer Mitigation over BAU [business-as-usual], given that massive human rights violations would occur either way. (p. 232)

Shue (2014, pp. 316–18) makes a similar point.

The root of the problem is that about 81 percent of the world's energy is produced by fossil fuels (IEA, 2017, p. 6). That energy is, among other things, crucial to the development that is reducing poverty, malnutrition, disease, and early death around the world. If to prevent future harm we reduce fossil fuel consumption deeply and quickly, then (since we can't build a renewable energy infrastructure overnight) we cause more harm now. But if we do not cut fossil fuel use deeply and quickly, then we are likely to harm even more people—once again, primarily the poor—over the coming centuries. Either way the transgression of human rights is monumental.

If such are our choices, then it is futile to insist that rights violations are simply forbidden. Violations are, however, more or less egregious, and in a choice among evils, we should choose the lesser. Mollendorf continues:

> Providing guidance for policy in the case of competing human rights claims could perhaps be accomplished if we were to pursue the course of action that is likely to lead to maximal satisfaction of human rights.... There are respectable accounts of rights that take them as the objects of maximizing strategies. (2014, p. 232)[9]

Though Mollendorf thus entertains the idea of a consequentialism of rights, he shrinks from endorsing it, because

> an important part of the moral importance of human rights is to offer protections to minorities against the demands of majorities, and it is at least questionable that a consequentialism of rights can adequately accomplish that. (2014, p. 232).

His worry is that a consequentialism of rights might license unjust distribution of rights satisfaction and/or violation.

It is not clear whether Mollendorf is speaking of *inter*generational or of *intra*generational distribution. But for long-term climate justice, it doesn't much matter, since we have no effective control over either. As always, it will be the poor of those generations whose rights are most cruelly violated.

[9] The "respectable account" that Mollendorf mentions in particular is that of Sen (1982).

But the degree to which their rights are respected by their contemporaries or their nearer predecessors is not up to us. We know that under harsher conditions there is likely to be more injustice, but we can't reliably influence either the distant *intra*generational distribution of that injustice or its distant *inter*generational distribution. A long-term consequentialism of rights need not, therefore, offer protections to future minorities against the demands of future majorities, either intergenerationally or intragenerationally, because it can't.[10]

What it can do (perhaps the *only* thing it can do) is to insist that we refrain, insofar as possible, from violating future people's rights to non-harm. A long-term consequentialism of rights is thus both feasible and conceptually fairly simple; what we owe to *distant* future people is simply not to harm them, if we can help it. We must, of course, balance the rights of distant future people against the rights of present people and near-future people, especially the poor. Hence we should retain a more robust human rights theory for the near term, where our ability to predict and control events is greater. But ultimately the presumption that human rights are, in every individual, morally inviolable must in the long term give way to an imperative to keep aggregate harm as low as we reasonably can, consistent with present responsibilities. Long-term climate ethics thus takes on a distinctly consequentialist hue.

13.5. A Consequentialism of Rights—or Just Plain Consequentialism?

By "consequentialism," I do not mean the attempt to maximize utility, if utility is defined as overall preference satisfaction or pleasure minus pain. The harms at issue here (injury, sickness, homelessness, and death) are losses of *objective* welfare; hence objective welfare is the relevant form of "utility." Objective welfare can be assessed by measures of public health, education, employment, safety, and the like.

Nor do I mean an imperative to *maximize* anything. Decisions at all levels, both public and private, typically are, and ought to be, made by considering a multidimensional array of competing and often incommensurable values. The available choices, when ranked in accord with such values, are usually merely partially ordered. There are better choices and worse, but often there is no choice that maximizes value. Decision theorists routinely impose a

[10] It is perhaps worth noting that it therefore need not protect minorities from the tyranny of the majority in *any* case, since we fossil fuel users are most likely a *minority*, rather than a majority, relative to the vast numbers of future people who will have to deal with our climate disruption.

maximum on raw value rankings by constructing a preference function over them (Keeny and Raiffa, 1993, ch. 3), but ethicists ought always to question the objectivity of that practice.

Ethical consequentialist decision procedures over partially ordered values typically require satisficing: eliminating the worst options and choosing, from among the better, one that is good enough. Objective welfare consequentialism satisfices over some "objective list" of values. I assume that a consequentialism of rights would proceed similarly, using one or more measures of rights violation or satisfaction. In the long-term case, the violations would be harms.

Yet, even in the long term, when rights violations are coextensive with an array of serious harms, plain consequentialism and consequentialisms of rights can differ. Plain consequentialism holds that even serious harms can be balanced by equally weighty benefits. A consequentialism of rights would hold that rights violations cannot be justified by their benefits, though it would permit some rights violations for the sake of avoiding weightier ones.

This leads some plain consequentialists (and especially some economists) to reject rights-based climate justice. Our primary responsibility, they argue, is to promote innovation and growth, whose benefits will outweigh the harms that all but the most devastating levels of climate change could produce. Given their optimism, plain consequentialism clearly diverges from a consequentialism of rights.

Other theorists (I am one) foresee a more tragic future—one in which economies bog down amid worldwide disruptions. Given such pessimism, plain consequentialism and a consequentialism of rights agree with respect to the long term: choose the lesser of evils. Both imply that we may not harm distant future people, just as we may not harm present people, without moral necessity. For both, what long-term climate justice requires is, first and foremost, strenuous reduction of greenhouse gas emissions. Thus whether plain objective welfare consequentialism and consequentialisms of rights yield similar duties in the long term depends largely upon whether long-term prospects are bright or bleak.

One final point: everything said here assumes a purely anthropocentric ethic. If non-human beings also matter, the prospects are undoubtedly bleak.

13.6. Climate Justice and Moral Psychology

There remains the problem of moral psychology. Though I argued earlier that *long-term* human rights theory lacks a viable moral psychology, this is not true of shorter-term human rights theory. The rhetorical, political, and moral effectiveness of rights discourse in current and historically recent struggles for justice is beyond question. Where that discourse has succeeded, however, it has often tapped more ancient and more universal wellsprings of human emotion.

"Let justice roll down like waters," cries the prophet Amos, "and righteousness like an ever-flowing stream" (Amos 5:24). Justice for Amos is neither fair distribution nor respect for rights, nor promotion of objective welfare. It is God's judgment against idolatry, arrogance, the heedlessness of the rich, and the oppression of the helpless by the powerful.

I am not waxing theological. We may think of Amos' God merely as the depths of the human psyche projected into the sky. But the power of his rhetoric reveals something important about human motivation. Injustice rouses God's anger—an anger fueled by sympathy with the powerless and ignited when the arrogant and powerful afflict them in pursuit of personal gain. These emotions are human and universal.

John Stuart Mill analyzes such feelings in chapter 5 of *Utilitarianism*. Nominally, the topic is justice itself, but the bulk of the chapter concerns "the sentiment of justice"—or, more generally, moral psychology. According to Mill, the sentiment of justice is

> the animal desire to repel or retaliate a hurt or damage to oneself, or to those with whom one sympathizes, widened so as to include all persons, by the human capacity of enlarged sympathy, and the human conception of intelligent self-interest. From the latter elements the feeling derives its morality; from the former, its peculiar impressiveness and energy of self-assertion. (1861, p. 65)

As in Amos, sympathy for the powerless fuels anger at the powerful who afflict them. Says Mill:

> The feelings concerned are so powerful, and we count so positively on finding a responsive feeling in others (all being alike interested), that *ought* and *should* grow into *must*, and recognized indispensability becomes a moral necessity, analogous to physical, and often not inferior to it in binding force. (p. 67)

Anger is, of course, treacherous. Mill rightly warns that "the natural feeling of retaliation or vengeance . . . has nothing moral in it." Yet it can be sublimated, beneficially, into moral action. "[W]hat is moral," Mill says, "is, the exclusive subordination of it to the social sympathies, so as to wait on and obey their call" (p. 64).

Mill is, of course, a plain consequentialist, but his moral psychology is not specific to consequentialism. Amos—for whom justice is interwoven with the same, or at least similar, emotions—has a divine-command ethic. The same moral psychology can be, and often has been, used in the service of human rights. I see no reason why it could not also be adapted to a consequentialism of rights.

But is it practical for long-term climate ethics? We are easily moved to anger when offenders harm us or people we care about. But in the case of climate justice, we and the people we most care about are harming others whom we will never know and who have no recourse against us. We may thus be moved less to anger at the affliction of the powerless by the powerful than to denial and derision.

Yet we are not all equally implicated in climate harm, and that inequality gives moral anger some social purchase. Many of us would happily switch from fossil fuels to renewable energy if they could. Some already have. Many conserve energy for moral reasons. They then naturally resent others who live in trophy houses, drive mammoth SUVs, deny climate science, or undercut sane efforts to reduce carbon emissions. Emotions such as these do in fact energize climate justice work today, just as they have long energized other social justice struggles. They are evident in climate demonstrations worldwide.[11]

There is, moreover, disunity within each of us, and that gives moral indignation some *psychological* purchase. We can be motivated by our natural revulsion at the spectacle of the powerful, heedless, and arrogant unjustly harming people who are powerless, voiceless, and without recourse against them—even if the powerful, heedless, and arrogant are us.

Acknowledgments

This chapter originated as a contribution to the conference on Climate Justice: Economics and Philosophy at Cornell University in May 2016 and was revised in response to comments from colleagues at the University of Tennessee and at the Center for Ethics and Social Values at Ohio State University. Thanks are due to Nolan Hatley, Trevor Hedberg, Don Hubin, Corey Katz, J. Paul Kelleher, Eden Lin, Tristram McPherson, Annette Mendola, Henry Shue, and Piers Turner for helpful criticisms, questions, and suggestions.

References

Archer, D. et al. 2009. “Atmospheric Lifetime of Fossil Fuel Carbon Dioxide.” *Annual Review of Earth and Planetary Sciences* 37: 117.

Broome, J. 2012. *Climate Matters: Ethics in a Warming World*. New York: W. W. Norton.

Caney, S. 2010. “Climate Change, Human Rights, and Moral Thresholds.” In *Human Rights and Climate Change*, edited by Stephen Humphreys, 69–90. Cambridge: Cambridge University Press.

[11] See, for example, <https://peoplesclimate.org/>.

DARA (Development Assistance Research Associates). 2012. *Climate Vulnerability Monitor*, 2nd edition. <http://daraint.org/climate-vulnerability-monitor/climate-vulnerability-monitor-2012/report/> accessed January 16, 2018.

Gardiner, S. 2003. "The Pure Intergenerational Problem." *Monist* 86 (3): 481.

Gardiner, S. 2009. "A Contract on Future Generations?" In *Intergenerational Justice*, edited by A. Gosseries and L. Meyer, 77–118. Oxford and New York: Oxford University Press.

Gardiner, S. 2011. *A Perfect Moral Storm: The Ethical Tragedy of Climate Change*. Oxford and New York: Oxford University Press.

Gleick, Peter H. 2014. "Water, Drought, Climate Change, and Conflict in Syria." *Weather, Climate, and Society* 6: 331.

Huebert, R., H. Exner-Pirot, A. Lajeunesse, and J. Gulledge. 2012. "Climate Change and International Security: The Arctic as a Bellwether." Center for Climate and Energy Solutions, Arlington, Virginia. <https://www.c2es.org/site/assets/uploads/2012/04/arctic-security-report.pdf> accessed January 16, 2018.

IAP (Interacademy Panel on International Issues). 2009. "IAP Statement on Ocean Acidification." <http://www.interacademies.net/File.aspx?id=9075> accessed January 16, 2018.

IEA (International Energy Agency). 2017. "Key World Energy Statistics 2017." <https://www.iea.org/publications/freepublications/publication/KeyWorld2017.pdf>, accessed May 2, 2018.

Kavka, G. 1982. "The Paradox of Future Individuals." *Philosophy and Public Affairs* 11 (2): 93.

Keeny, R. and H. Raiffa. 1993. *Decisions with Multiple Objectives: Preferences and Value Tradeoffs*. Cambridge: Cambridge University Press.

Maudlin, T. 2011. *Quantum Non-Locality and Relativity*, 3rd edition. Malden, MA: Wiley-Blackwell.

Meyer, L. 2015. "Intergenerational Justice." *Stanford Encyclopedia of Philosophy*. <https://plato.stanford.edu/entries/justice-intergenerational/> accessed January 16, 2018.

Mill, J. S. 1861. *Utilitarianism*. Indianapolis: Bobbs-Merrill.

Mollendorf, D. 2014. *The Moral Challenge of Dangerous Climate Change*. Cambridge: Cambridge University Press.

Mozersky, J. 2011. Presentism. In *The Oxford Handbook of Philosophy of Time*, edited by C. Callendar, 122–44. Oxford and New York: Oxford University Press.

Nolt, J. 2011a. "Greenhouse Gas Emission and the Domination of Posterity." In *The Ethics of Global Climate Change*, edited by D. Arnold, 60–76. Cambridge: Cambridge University Press.

Nolt, J. 2011b. "How Harmful Are the Average American's Greenhouse Gas Emissions?" *Ethics, Policy and Environment* 14: 3.

Nolt, J. 2013. Replies to Critics of "How Harmful Are the Average American's Greenhouse Gas Emissions?" *Ethics, Policy and Environment* 16 (1): 111.

Nolt, J. 2015a. *Environmental Ethics for the Long Term*. New York: Routledge.

Nolt, J. 2015b. "Casualties as a Moral Measure of Climate Change." *Climatic Change* 130 (3): 347.

Nolt, J. Forthcoming a. "Domination across Space and Time: Smallpox, Relativity, and Climate Ethics." *Ethics, Policy and Environment.*

Nolt, J. Forthcoming b. "Cumulative Harm as a Function of Carbon Emissions." In *Climate Change and its Impacts: Risks and Inequalites*, edited by C. Murphy, P. Gardoni and R. McKim. Cham, Switzerland: Springer.

Parfit, D. 1984. *Reasons and Persons*. Oxford and New York: Oxford University Press.

Purves, D. 2016. "The Case for Discounting the Future." *Ethics, Policy & Environment* 19 (2): 213.

Savitt, S. 2011. "Time in the Special Theory of Relativity." In *The Oxford Handbook of Philosophy of Time*, edited by C. Callendar, 546–70. Oxford and New York: Oxford University Press.

Sen, A. 1982. "Rights and Agency." *Philosophy and Public Affairs* 11: 3.

Shue, H. 2014. *Climate Justice: Vulnerability and Protection*. Oxford and New York: Oxford University Press.

Springman, M. et al. 2016. "Global and Regional Health Effects of Future Food Production under Climate Change: A Modeling Study." *The Lancet* 387: 1937.

Vanderheiden, S. 2008. *Atmospheric Justice: A Political Theory of Climate Change*. Oxford and New York: Oxford University Press.

WHO (World Health Organization). 2014. "Quantitative Risk Assessment of the Effects of Climate Change on Selected Causes of Death, 2030s and 2050s." <http://www.who.int/globalchange/publications/quantitative-risk-assessment/en/> accessed January 16, 2018.

Woodward, J. 1986. "The Non-Identity Problem." *Ethics* 96 (4): 804.

Zeebe, R. 2013. "Time-Dependent Climate Sensitivity and the Legacy of Anthropogenic Greenhouse Gas Emissions." *Proceedings of the National Academy of Sciences* 110: 13739.

Zimmerman, D. 2011. "Presentism and the Spacetime Manifold." In *The Oxford Handbook of Philosophy of Time*, edited by C. Callendar, 162–244. Oxford and New York: Oxford University Press.

Appendix

In the run up to Paris, Mary Robinson instituted the Climate Justice Dialogue. As part of this project, the global High Level Advisory Committee to the Climate Justice Dialogue was formed in 2013. The committee included former presidents and other leaders from the fields of politics, science, business, civil society, and academia, and publicly advocated for climate justice through their own work and via platforms facilitated by the Mary Robinson Foundation. Together, they signed the Declaration on Climate Justice, which is reproduced here.[1]

Declaration on Climate Justice

18 September 2013

> "All human beings are born free and equal in dignity and rights."[2]

Our vision

As a diverse group of concerned world citizens and advocates, we stand in defence of a global climate system that is safe for all of humanity. We demand a world where our children and future generations are assured of fair and just opportunities for social stability, employment, a healthy planet and prosperity.

We are united in the need for an urgent response to the climate crisis—a response informed by the current impacts of climate change and the science that points to the possibility of a global temperature increase of 4°C by the end of this century. The economic and social costs of climate impacts on people, their rights, their homes, their food security and the ecosystems on which they depend cannot be ignored any longer. Nor can we overlook the injustice faced by the poorest and most vulnerable who bear a disproportionate burden from the impacts of climate change.

This reality drives our vision of climate justice. It puts people at the centre and delivers results for the climate, for human rights, and for development. Our vision acknowledges the injustices caused by climate change and the responsibility of those who have caused it. It requires us to build a common future based on justice for those

[1] <https://www.mrfcj.org/wp-content/uploads/2015/09/Declaration-on-Climate-Justice.pdf>.

[2] Article 1, The Universal Declaration of Human Rights.

who are most vulnerable to the impacts of climate change and a just transition to a safe and secure society and planet for everyone.

Achieving climate justice

A greater imagination of the possible is vital to achieve a just and sustainable world. The priority pathways to achieve climate justice are:

Giving voice: The world cannot respond adequately to climate change unless people and communities are at the centre of decision-making at all levels—local, national and international. By sharing their knowledge, communities can take the lead in shaping effective solutions. We will only succeed if we give voice to those most affected, listen to their solutions, and empower them to act.

A new way to grow: There is a global limit to the carbon we can emit while maintaining a safe climate and it is essential that equitable ways to limit these emissions are achieved. Transforming our economic system to one based on low-carbon production and consumption can create inclusive sustainable development and reduce inequality. As a global community, we must innovate now to enable us to leave the majority of the remaining fossil fuel reserves in the ground—driving our transition to a climate resilient future.

To achieve a just transition, it is crucial that we invest in social protection, enhance worker's skills for redeployment in a low-carbon economy and promote access to sustainable development for all. Access to sustainable energy for the poorest is fundamental to making this transition fair and to achieving the right to development. Climate justice also means free worldwide access to breakthrough technologies for the transition to sustainability, for example, efficient organic solar panels and new chemical energy storage schemes.

Investing in the future: A new investment model is required to deal with the risks posed by climate change—now and in the future, so that intergenerational equity can be achieved. Policy certainty sends signals to invest in the right things. By avoiding investment in high-carbon assets that become obsolete, and prioritizing sustainable alternatives, we create a new investment model that builds capacity and resilience while lowering emissions.

Citizens are entitled to have a say in how their savings, such as pensions, are invested to achieve the climate future they want. It is critical that companies fulfil their social compact to invest in ways that benefit communities and the environment. Political leaders have to provide clear signals to business and investors that an equitable low-carbon economic future is the only sustainable option.

Commitment and accountability: Achieving climate justice requires that broader issues of inequality and weak governance are addressed both within countries and at a global level. Accountability is key. It is imperative that Governments commit to bold action informed by science, and deliver on commitments made in the climate change regime to reduce emissions and provide climate finance, in particular for the most vulnerable countries.

All countries are part of the solution but developed countries must take the lead, followed by those less developed, but with the capacity to act. Climate justice increases the likelihood of strong commitments being made as all countries need to be treated fairly to play their part in a global deal. For many communities, including indigenous peoples around the world, adaptation to climate change is an urgent priority that has to be addressed much more assertively than before.

Rule of law: Climate change will exacerbate the vulnerability of urban and rural communities already suffering from unequal protection from the law. In the absence of adequate climate action there will be increased litigation by communities, companies and countries. International and national legal processes and systems will need to evolve and be used more imaginatively to ensure accountability and justice. Strong legal frameworks can provide certainty to ensure transparency, longevity, credibility and effective enforcement of climate and related policies.

Transformative leadership

World leaders have an opportunity and responsibility to demonstrate that they understand the urgency of the problem and the need to find equitable solutions now.

At the international level and through the United Nations, it is crucial that leaders focus attention on climate change as an issue of justice, global development and human security. By treating people and countries fairly, climate justice can help to deliver a strong, legally binding climate agreement in 2015. It is the responsibility of leaders to ensure that the post-2015 development agenda and the UNFCCC climate negotiations support each other to deliver a fair and ambitious global framework by the end of 2015. Local and national leaders will implement these policies on the ground, creating an understanding of the shared challenge amongst the citizens of the world and facilitating a transformation to a sustainable global society.

As part of global collective action, greater emphasis should be given to the role of diverse coalitions that are already emerging at the community, local, city, corporate and country levels and the vital role they play in mobilizing action. These coalitions are already championing the solutions needed to solve the crisis and their effect can be maximized by supporting them to connect and scale up for greater impact.

Climate justice places people at its centre and focuses attention on rights, opportunities and fairness. For the sake of those affected by climate impacts now and in the future, we have no more time to waste. The 'fierce urgency of now' compels us to act.

18 September 2013

This Declaration is supported by:

Mr Nnimmo Bassey, Coordinator, Oilwatch International

Ms Sharan Burrow, General Secretary, International Trade Union Confederation

Ms Luisa Dias Diogo, Former Prime Minister, Mozambique

Ms Patricia Espinosa-Cantellano, Ambassador of Mexico to Germany

Mr Bharrat Jagdeo, Roving Ambassador for the Three Basins Initiative, Former President of Guyana

Prof Pan Jiahua, Director General, Institute for Urban and Environmental Studies, Chinese Academy of Social Sciences

Prof Ravi Kanbur, TH Lee Professor of World Affairs, International Professor of Applied Economics and Management, Professor of Economics, Cornell University

Mr Caio Koch-Weser, Vice Chairman, Deutsche Bank Group, Chairman of the Board, European Climate Foundation

Mr Ricardo Lagos, President, Fundacion Democracia y Desarrollo, Former President of Chile

Mr Festus Mogae, Member, African Union High-level Panel for Egypt, Former President of Botswana

Mr Jay Naidoo, Chair of the Board of Directors, Chair of the Partnership Council, Global Alliance for Improved Nutrition

Mr Marvin Nala, Campaigner, Greenpeace East Asia

Prof Kirit Parikh, Chairman, Integrated Research and Action for Development, IRADe

Ms Sheela Patel, Founder-Director, Society for the Promotion of Area Resource Centers

Mrs Mary Robinson, President, Mary Robinson Foundation—Climate Justice, Former President of Ireland

Prof Hans Joachim Schellnhuber, Director, Potsdam Institute for Climate Impact Research

Prof Henry Shue, Senior Research Fellow at Merton College and Professor of Politics and International Relations, University of Oxford

Mr Tuiloma Neroni Slade, Secretary General, Pacific Islands Forum Secretariat

Dr Andrew Steer, President and CEO, World Resources Institute

Ms Victoria Tauli-Corpuz, Executive Director, Tebtebba Foundation

Ms Dessima Williams, Former Ambassador of Grenada to the United Nations

Those listed above are members of the High Level Advisory Committee (HLAC) to the Climate Justice Dialogue, an initiative of the Mary Robinson Foundation—Climate Justice and the World Resources Institute, that aims to mobilize political will and creative thinking to shape an ambitious and just international climate agreement in 2015.

Index